Science Experiments by the Hundreds

Second Edition

Authors in alphabetical order:

Julia H. Cothron
Mathematics & Science Center
Richmond, Virginia

Ronald N. Giese
The College of William and Mary
Williamsburg, Virginia

Richard J. Rezba
Virginia Commonwealth University
Richmond, Virginia

KENDALL/HUNT PUBLISHING
4050 Westmark Drive

Disclaimer

Adult supervision is required when working on projects. Use proper equipment (gloves, forceps, safety glasses, etc.) and take other safety precautions such as tying up loose hair and clothing and washing your hands when the work is done. Use extra care with chemicals, dry ice, boiling water, or any heating elements. Hazardous chemicals and live cultures (organisms) must be handled and disposed of according to appropriate directions from your adult advisor. Follow your science fair's rules and regulations and the standard scientific practices and procedures required by your school. No responsibility is implied or taken for anyone who sustains injuries as a result of using the materials or ideas, or performing the procedures described in this book.

Additional safety precautions and warnings are mentioned throughout the text. If you use common sense and make safety a first consideration, you will create a safe, fun, educational, and rewarding project.

Book Team

Chairman and Chief Executive Officer Mark C. Falb
Director of National Book Program Paul B. Carty
Editorial Development Manager Georgia Botsford
Developmental Editor Tina Bower
Assistant Vice President, Production Services Christine E. O'Brien
Prepress Project Coordinator Angela Shaffer
Permissions Editor Renae Heacock
Design Manager Deb Howes
Senior Vice President, College Division Thomas W. Gantz
Managing Editor, College Field Paul Gormley
Associate Editor, College Field Curtis Ross

Cover photos © PhotoDisc, Inc. and Corel

© 1996, 2004 Kendall/Hunt Publishing Company

ISBN 0-7575-0971-1

All rights reserved. No part of this publication may be reproduced, stored in a retrieval system, or transmitted, in any form or by any means, electronic, mechanical, photocopying, recording or otherwise, without the prior written permission of the copyright owner.

Printed in the United States of America

10 9 8 7 6 5 4 3 2

Contents

Letter to Students v

1. ◆ **Conducting Experiments** 1
   Variables, Constants, and Hypotheses

2. ◆ **Refining Experiments** 9
   Controls, Data, and Repeated Trials

3. ◆ **Analyzing Experiments** 25
   Diagrams, Flaws, and Improvements

4. ◆ **Finding Project Ideas** 37
   Interests, Labs, Books, and Magazines

5. ◆ **Brainstorming Project Ideas** 47
   Four Question Strategy

6. ◆ **Exploring a Project Idea** 59
   Libraries, Note Cards, and Interviews

7. ◆ **Writing Procedures** 75
   Materials, Steps, and Paragraphs

8. ◆ **Experimenting Safely** 87
   Organisms, Chemicals, Shocks, and Radiation

9. ◆ **Recording Data** 95
   Tables and Data Summaries

10. ◆ **Constructing Bar Graphs** 109
    Axes, Bars, Distributions, and Sentences

11. ◆ **Constructing Line Graphs** 125
    Plots, Lines, and Trends

12. ◆ **Writing an Experimental Report** 137
    From Introduction to Conclusion

13. ◆ **Making Stem-and-Leaf Plots** 147
    Data Displays and Distribution Patterns

14. ◆ **Making Boxplots or Box and Whisker Diagrams** 161
    Graphic Displays, Quartiles, and Data Spread

15. ◆ **Presenting Your Experiment** 177
    Rules, Displays, Posters, and Presentations

Glossary 191

Appendix A: Using Style Manuals 196

Appendix B: Conducting Interviews 199

Appendix C: Expanding an Introduction 201

You just performed an experiment. In an experiment, you change something to see what happens. Things in an experiment that change are called **variables**. Here, you changed the temperature of ketchup. You made this change on purpose to learn how temperature affects the flow of ketchup.

Independent Variable

In the ketchup experiment you purposely changed the temperature of the ketchup. Scientists call the variable that is purposely changed the **independent** or **manipulated variable**. In the experiment you have one independent variable—the temperature of the ketchup. You changed the ketchup temperature three ways by using ice, room temperature, and hot water. The three ways that you changed the ketchup temperature are called the **levels of the independent variable**.

When designing an experiment as a beginning researcher, you choose only one variable that you purposely change. Why? If you change more than one variable, such as temperature *and* brand of ketchup, you may never know which variable caused the response.

Dependent Variable

In the ketchup experiment there was also a second variable—the distance the ketchup flowed in 30 seconds. The ketchup flowed different distances because the blobs were different temperatures. The variable that responds in an experiment is the **dependent** or **responding variable.**

Constants

Certain things were kept the same in the ketchup experiment. You used the same brand of ketchup. The ketchup was placed on the same type of plate and tilted the same amount. The travel time was also the same for all temperatures of ketchup. Things that are kept the same in an experiment are called **constants**. Can you think of other constants in the ketchup experiment? If your results were not what you expected, perhaps another variable in addition to ketchup temperature was affecting how the ketchup moved.

Constructing Hypotheses

Before you conduct an experiment, you should try to predict what is going to happen. You need to think about how changing your independent variable will affect your dependent variable. This prediction is called a **hypothesis**. An example of a hypothesis for the ketchup experiment is: **If** the cost of the ketchup increases, **then** the time to flow decreases. To write such a hypothesis, use an "**If...**, **then**" sentence: **If** the (independent variable) is (describe how you changed it), **then** the (dependent variable) will (describe the effect). For *Investigation 1.1, The Great Tomato Race*, a hypothesis might be:

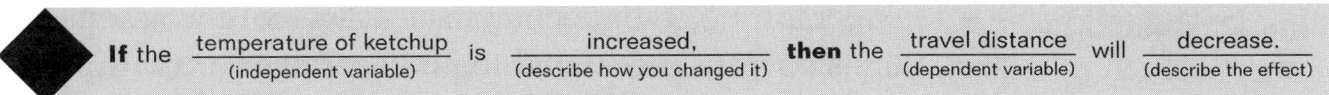

When scientists conduct an experiment, they often make a hypothesis about what they think will happen. A hypothesis is a prediction that is not a wild guess. Most hypotheses are based on careful study. Hypotheses are based on observations, previous experimental results, and information from books and meetings with other scientists.

Support for the Hypothesis

When the hypothesis and the experimental results agree, scientists say the hypothesis was **supported** by the results. When the hypothesis and the results do not agree, the hypothesis is **not supported**. Scientists do not say a hypothesis was right or wrong, correct or incorrect. Instead, we say that our hypothesis was either supported, or not supported, by the experiment results. In the ketchup experiment did you think about which temperature of ketchup would win the race? Did your results support or not support your hypothesis? To prevent ketchup from dripping out of a hamburger, which side of the bun should you put the ketchup on—the hot meat side or the cooler lettuce side?

Scientists do not say they were right or wrong or that their hypothesis was correct or incorrect, and neither should you. Similarly, scientists do not say that a single experiment proves or disproves a hypothesis. Each experiment provides evidence that a hypothesis is supported or not supported. Many experiments must be conducted before results are accepted as fact. For example, you would need to do the experiment many times to determine the effect of temperature on ketchup flow.

Conducting Experiments **3**

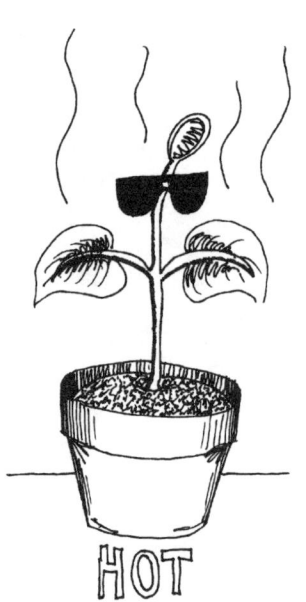

You would need to do even more experiments if you wanted to know how temperature affected other substances such as mustard, mayonnaise, and steak sauce.

Hot and Cold Plants

In ***Investigation 1.1, The Great Tomato Race***, you learned how temperature affected the flow of ketchup. What about living things? How would house plants be affected by hot and cold? Suppose your best friend's family grew African violets as a hobby. If you were helping him water those plants, what temperature of water should be used? If you and your friend read books about plants, or the instructions that came with the plants, you would find some information. By reading these materials both of you would probably learn about ways to water plants, amount of water to use, and how often to water. But what if there was no information on the proper temperature of water? Could you conduct an experiment to find out? What would be your hypothesis?

African Violets are expensive, so here's an experiment to learn about the effect of temperature on plants using a less costly plant, such as some variety of string bean.

> Plant four bean seeds in identical pots containing the same amount and kind of soil. Label the pots A, B, C, and D and place them in the same sunny window. On Mondays and Fridays, water each plant with 125 mL of water at the temperatures shown in Table 1.1. At the end of 30 days measure the height of the plants in centimeters.

After reading the description of the plant experiment, can you identify the independent variable? the dependent variable? the constants?

In this experiment, one independent variable is used—the temperature of water. The water temperature is changed four ways by using ice, room temperature, hot, and boiling water. The four ways that the water temperature is changed are the levels of the independent variable.

4 Chapter 1

In the bean experiment there is also a second variable—the height of plants. The plants grow different amounts because they receive different temperatures of water. In this experiment the height of the plants is the dependent variable.

Certain things are kept the same in the experiment. The same type of plant is used—beans. The plants also receive the same amount of water—125 mL. The things that should be kept the same are the constants. Other constants in this experiment are the size of the pot, type of soil, and location in a sunny window.

What hypothesis did you make for the experiment? One hypothesis could be: **If** the temperature of water is increased, **then** the plants will grow less. Here's the data we collected.

TABLE 1.1 The Effect of Water Temperature on the Height of Bean Plants

Bean Plant	Temperature of Water	Height of Bean Plant (cm)
A	Room temperature	24.0
B	Hot tap water	18.0
C	Ice water	22.0
D	Boiling water	0.0

Did the data support the hypothesis? What about the hypothesis you made?

Which soaps around your house make the most suds? Do you think changing the type of soap (dish detergent, dishwasher powder, bar soap) would affect the amount of suds produced? Now that you know something about experiments, use your new knowledge to make a hypothesis and conduct an experiment to test it.

Investigation 1.2

A Sudsational Experience

 **WHAT YOU NEED**

- 3 100 mL graduated cylinders
- Plastic wrap
- 3 rubber bands
- Metric measuring spoon
- Different soaps (dish detergent, dishwasher powder, bar soap)
- Safety goggles

 **WHAT YOU LEARNED**

1. What was the variable you purposely changed in the experiment (independent variable)?
2. What was the variable that responded (dependent variable)?
3. What were the things you kept the same (constants)?
4. What was your hypothesis? Was it supported by the data?
5. How could you improve the experiment?

 **WHAT YOU DO**

1. Fill each graduated cylinder with 50 mL of water.
2. Add 1 mL of a different soap to each graduated cylinder.
3. Seal the top of each graduated cylinder with a piece of plastic wrap and put a rubber band around the wrap to hold it tight. Shake each graduated cylinder for 15 sec.
4. Measure in milliliters the volume of the suds produced above the water line. Record your results.
5. Pour out the solutions and rinse the graduated cylinders.

Take care when using rubber bands. See Chapter 8, Experimenting Safely, Section A, Chemicals.

6 Chapter 1

Searching the Web

"Searching the Web" boxes appear in every chapter. These special boxes provide one website address as well as words and phrases you can use to search the Internet for additional information about the chapter.

Follow this example to narrow a search that began with **Science Fair Projects** and resulted in 98,900 web page listings.

    Add **"middle school"** This narrowed it to 8,200 web page listings.
    Add **"physical science"** This narrowed it to 1,270 web page listings.
    Add **magnets** This narrowed it to 141 web page listings, a much more manageable number of web pages to choose from.

For additional information see "Locating Digital Information" in Chapter 6.

For more information on Conducting Experiments, use these search words or phrases:

variables
constants
hypothesis
"designing experiments"
"parts of an experiment"
"components of an experiment"

to find websites such as this:
longwood.edu/cleanva/images/sec6.designexperiment.pdf

Practice Problems

Conducting Experiments

1. Identify the independent variable and the dependent variable.
   a. Michael put 100 red seeds, 100 brown seeds, and 100 yellow seeds in his bird feeder. He counted the number of seeds of each color that remained after 2 days.
   b. Leticia timed how fast apple slices turned brown after being dipped in different preservatives, such as lemon juice, fruit freshener, and lime soda.

2. Identify the independent variable, dependent variable, and constants.
   a. Elizabeth tested how high 5 different brands of new tennis balls would bounce when dropped from a height of 2 meters. She dropped the balls so that they hit the same floor tile each time.
   b. Maria wondered if different colors of plastic wrap affected the time for bread to mold. She wrapped a slice of bread in clear plastic wrap. Before wrapping the bread she placed 5 mL of water in the center of the bread. She repeated the process with a slice of bread using red plastic wrap and a slice of bread using blue plastic wrap. The bread came from the center of the same loaf. In each case she used the same amount and same brand of plastic wrap. She wrapped the different slices of bread in the same way.

3. Identify the independent variable and dependent variable in each hypothesis.
   a. If liquids are placed in containers with sides of different heights, then they will evaporate faster in the container with the lower sides.
   b. If a greater amount of soap is added to water, then fewer drops of water can be placed on a penny.

4. Determine if the hypothesis is supported by the data.
   Katherine thought that people would be able to taste small amounts of sugar more easily than vinegar or salt. She found that people could detect 2 mL of salt in a liter of water. However, they needed only 1.5 mL of sugar per liter of water and 0.5 mL of vinegar per liter of water to taste the substance.

5. Make a hypothesis about the following:
   How hard rock, rock and roll, and classical music affect pulse rate

8 Chapter 1

Refining Experiments

Controls, Data, and Repeated Trials

Suppose you and your best friend decided to bake a surprise birthday cake for your grandmother. Talk about experiments! The recipe calls for two eggs. When your friend takes the eggs out of the refrigerator, she mutters, "I wonder how old these eggs are?" You watch in amazement as she puts the eggs in a bowl of water, observes them sinking to the bottom, and says: "These are okay to use." When questioned, your friend explains, "Fresh eggs sink in water. Eggs that float are so old that they have spoiled."

Now consider this: "Objects float higher in salt water than in fresh water because salt water is denser. In fact, some things that sink in fresh water will float in salt water." What about fresh eggs? Will they float in salt water? Try **Investigation 2.1: Can an Egg Float?** to find out.

Investigation 2.1

Can an Egg Float?

 **WHAT YOU NEED**

- Safety goggles
- 1 uncooked egg in the shell
- Table salt
- A tall clear plastic or glass container (10–12 oz.)
- Warm water
- Metric ruler
- Container to keep egg from rolling away
- Graduated cylinder
- 5 mL Metric measuring spoon

See Chapter 8, Experimenting Safely, Section A, Chemicals.

 **WHAT YOU DO**

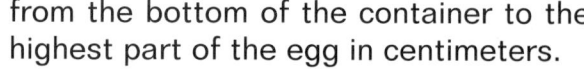

1. Fill a container with 250 mL of warm water.
2. Put the egg into the water and measure the distance from the bottom of the container to the highest part of the egg in centimeters.
3. Remove the egg and place it in the container.
4. Add 5 mL of salt to the water and stir until the salt is dissolved.
5. Gently put the egg in the salt water and measure the highest part of the egg.
6. Repeat Steps 4 and 5 using a total of 10, 15, and 20 mL of salt.
7. Remove the egg. Discard the salt solution. Rinse the container thoroughly and repeat Steps 1 - 6 three more times for a total of four trials if time allows. Or combine your data with others in your class in a class data table.

Amount of Salt (mL)	Height of Top of Eggs (cm) Trials			
	1	2	3	4
0				
5				
10				
15				
20				

Now, check the sample data on page 11. How does your data compare to ours?

10   Chapter 2

TABLE 2.1 Sample Data for *Can an Egg Float?*

Amount of Salt (mL)	Height of Top of Eggs (cm) Trials			
	1	2	3	4
0	4	4	4	4
5	5	5	5	5
10	5	6	5	5
15	6	6	6	6
20	8	8	8	8

In Chapter 1 you learned several of the major parts of an experiment. Test your skills by identifying the variable you purposely changed (independent variable), the variable that responded (dependent variable), and the factors you kept the same (constants) in the experiment. What was your hypothesis? Did your data support your hypothesis?

Remember

The **independent variable** is purposely changed.

The **dependent variable** is the variable that responds.

The **constants** are the variables you keep the same.

There are several more important parts of an experiment that we have not discussed yet. Knowing these additional parts of an experiment will help you design even better experiments. Let's begin by thinking about several things you did in Investigation 2.1, *Can an Egg Float?*

1. Given that you knew that fresh eggs would not float in fresh water, why did you need to test "0 mL" of salt as part of the experiment?

2. Given that the answer to the question *Can an egg float?* is either yes or no, why did you measure the height the eggs floated?

3. Why was it necessary to repeat the investigation four times or combine the data from several groups working on the same problem?

From reading this chapter, you will find answering these questions as easy as floating in salt water.

Refining Experiments 11

Control Group

In Chapter 1 you learned that all of the factors that can affect the results must be kept constant except for the variable that you purposely change. When the independent variable is not changed, the dependent variable should not change. If the independent variable is unchanged and the dependent variable changes considerably, this indicates that something that should be a constant is changing. Experiments need a way to detect "hidden variables," that is factors that should be kept constant but which change accidentally. In the egg investigation, you needed a way to detect if some factor other than amount of salt was affecting the results. This is the reason for the plain tap water in the egg investigation. The trials involving just plain water are called **controls**. A control is used as a standard of comparison. In an experiment, a control is important because it is used to detect "hidden variables" that are varying when they should not.

In the egg investigation, the plain water was used as a standard. The changing heights of the egg in various concentrations of salt water were compared to the height in plain water to find out if the added salt affected the egg's flotation behavior. Each time salt was added, the egg floated higher; therefore, the amount of salt did affect flotation. Most experiments include a **control group**. All other experimental groups are compared with the control group to determine experimental effects. In some experiments, the control is called a **no treatment control**. In the egg investigation, the plain water was a no treatment control because zero or no amount of salt was added.

In some experiments, all trials receive a treatment. The experimenter must then select one of the levels of the independent variable being tested as the control to serve as the standard of comparison. In Chapter 1, you examined the effect of water temperature on plant growth. You studied room temperature, hot, ice, and boiling water. Giving plants "no temperature" water does not make sense—they would die—so a "no treatment" control won't work. Instead, you must select the plant watered with a certain temperature water as the control, and then state the reasons for your choice. For example, you could select the plants receiving room temperature water as the control. You could use the directions on the seed package as the explanation for your choice. This kind of a control is called an **experimenter selected control**.

Now you can probably answer the first thought question: "Given that you know that eggs would not float in fresh water, then why did you need to test 0 mL of salt as part of the experiment?" The answer is that testing the 0 mL of salt was needed as a control or standard of comparison. Before you can determine the effect of putting salt into water, you need to know the answer to several questions. "Are the eggs in your house fresh enough to sink?" "Does your egg behave the same as the egg in our investigation?" "Water varies from place to place—how do your eggs act in your water?" If the egg acts the same way each time it is put into plain water, you can be sure that any effects that you observe as you add salt result from the different amounts of salt added, and not to something else. Zero milliliters of salt is your no treatment control or standard of comparison.

Measurements and Counts as Types of Data

Measurements

Information collected in an experiment is called **data**. In the egg investigation you measured the distance from the bottom of the container to the highest part of the egg. To measure, you used a standard measuring instrument such as a ruler marked in millimeters. For everyday tasks we often use the English System of measurement. Examples of English units are inches, teaspoons, cups, quarts, pounds, and degrees Fahrenheit. Scientists, however, use the Metric System. Examples of metric units are millimeters, meters, liters, grams, and degrees Celsius. Because this book is written for use in both schools and home, your teacher may substitute English units of measurement for some activities you may be assigned to do at home. When metric measuring instruments are available, use the Metric System.

Counts

Data may also be collected by counting. Examples include the number of seeds that sprouted or the number of flowers produced.

In Investigation 2.1, the 0, 5, 10, 15, and 20 mL levels of the independent variable were tested. You may wish to extend the investigation using 20, 30, 40, 50, and 60 mL. In every experiment only a limited number of levels are tested. This frequently means that other interesting behaviors are not observed, but then no one experiment tests all the possible values of an independent variable.

By now you should be able to answer question 2: "Given that the answer to the question *Can an Egg Float?* is either yes or no, why did you measure the height the eggs floated?" A yes or no answer tells you a little bit. Collecting measure-

TABLE 2.2. The Effect of the Amount of Salt on the Height an Egg Floats

Amount of Salt (mL)	Height of Top of Eggs (cm) Trials				Mean Height of Egg (cm)
	1	2	3	4	
0	4	4	4	4	4
1	5	5	4	5	5
2	6	6	5	5	6
3	6	5	6	6	6
4	8	8	8	8	8

Investigation 2.1, *Can an Egg Float?* could have been done using a different responding variable (dependent variable). In addition to measuring The Effect of the Amount of Salt on the *Height an Egg Floats*, we also could have studied The Effect of the Amount of Salt on the *Tilt of an Egg*. Instead of measuring the height of the top of the egg as you did in Investigation 2.1, you would describe the tilt of the egg using a scale to describe how much the egg is tilted. Such a scale might look like this.

H = Horizontal T = Tilted
ST = Slight tilt V = Vertical

Notice that this time in the egg experiment the data are descriptions or observations rather than measurements.

Observations—A Type of Data

Although most often data are some type of measurement, another important type of data is observations. In an experiment with plants, you could record the color of the leaves as brown, yellow, light green, or dark green. Similarly, you could describe the stems as very sturdy, moderately sturdy, or not sturdy (limp). Observations are word descriptions of things such as color, cloudiness of water, or straightness of a path.

Sometimes you must observe the behavior of an object before you can make categories. For observations about the position of the egg in various concentrations of salt, we made a scale to show the tilt of the egg.

H = Horizontal
ST = Slight tilt
T = Tilted
V = Vertical

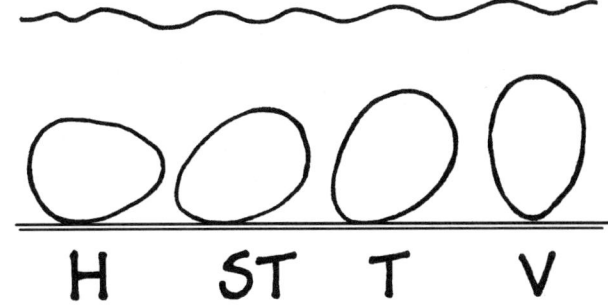

When you make a scale for your observations, you have to define what you mean by the categories. You can use words, diagrams, or a combination of both.

Observing the egg's position gives you additional information not contained in your measurements. Your measurements and observations give you two kinds of information about the dependent variable. Which is more important? That depends on what use you will make of the data. But two things are certain—if you don't collect the data while you're doing the experiment, you can not use it later if you need it, and if you do not collect the data, you're sure to need it later.

Summarizing Observations of Data

When your data are observations, such as the color of plants or the tilt of an egg, you cannot find the mean. Instead, you report the most frequent observation called the mode. The mode is another way to find the typical or central value. In the floating egg investigation, you would need to observe the position of the egg. Using the egg observation scale previously described, our data were recorded as shown in Table 2.3.

TABLE 2.3 The Effect of the Amount of Salt in a Solution on the Tilt of an Egg

Amount of Salt (mL)	Tilt of Egg Trials 1	2	3	4	Mode
0	H	H	H	T	H
5	T	T	H	T	T
10	T	T	T	V	T
15	V	V	V	V	V
20	V	V	V	V	V

Key: H = Horizontal   ST = Slightly titled   T = Tilted   V = Vertical

Refining Experiments   **17**

When you did the egg investigation, your results may have differed from ours. As you now know, the eggs, water, and salt you used might have been different from ours. In the next chapter we'll consider several ways to improve the procedure of an experiment so that the results vary less.

But first, read the following description of an experiment. Then see how quickly you can identify the:

- independent variable
- levels of the independent variable
- dependent variable
- number of repeated trials
- constants
- control

The **independent variable** is purposely changed.

The **dependent variable** is the variable that responds.

The **constants** are the variables you keep the same.

The **control** is a standard of comparison.

Keep on Trucking

Two students in a technology class were trying to find out if the amount of cargo affects the time that a battery operated truck takes to travel across a room. They hypothesized: **If** the number of cans of cargo is increased, **then** the truck's travel time will increase. In a series of truck races, they used different numbers of cans of soda (0, 1, 2, 3, and 4 cans) as cargo. They measured how many seconds the truck took to travel across the classroom floor with the different number of cans. They were careful to use the same truck, to place the truck in the same place on the floor, to aim the

truck in exactly the same way, and to remove all obstacles from the truck's path. They ran the sequence of races three times. Table 2.4 contains the data they collected. Does the data support the hypothesis?

TABLE 2.4 The Effect of the Amount of Cargo on the Travel Time of a Toy Truck

Cargo (Number of Cans)	Travel Time (sec) Trials 1	2	3	Mean Travel Time (sec)
0	11	13	12	12
1	20	21	25	22
2	34	34	28	32
3	42	40	42	41
4	54	50	50	52

The students were trying to determine the effect of the amount of cargo on the travel time of the battery operated truck. In the experiment, the amount of cargo was the independent or manipulated variable. Five levels for the independent variable were used—0, 1, 2, 3, and 4 cans. The responding or dependent variable was the time required, in seconds, for the truck to cross the floor. Constants were the same type of truck and path.

In this experiment, the empty truck serves as the standard of comparison or **control**. In an experiment, a control is important because it is used to detect factors that are varying when they should be kept constant. For example, suppose that each time the empty truck was raced, it went slower. This could mean the batteries were weakening, or that the wheels needed oil. What if sometimes the empty truck ran faster and sometimes slower? This could mean that the classroom floor was uneven or that there was a short circuit in the motor. Once the data collected for the empty truck (control) show that there were no such variations, the travel times with cans of soda can be compared with the empty truck's travel time to find out if the added cargo affected the truck's travel time. Each time a can was added, it took longer for the truck to cross the classroom floor. Having a control helps you to see the effects of the independent variable on the dependent variable. In this experiment, the empty truck is called a no treatment control because the empty truck received no amount of the independent variable.

In this experiment, there could be small chance variations in the roughness of the floor, or in the path the truck took across the floor, or in the position of the cans in the truck, or in the measurement of time. These variations could produce small chance errors in the measurements. To minimize these chance variations, repeated trials were conducted by measuring the truck's travel time three times with each load of 0, 1, 2, 3, or 4 cans. Repeated trials are the number of times each level of the independent variable is tested. In the truck experiment, there were five levels of the independent variable—0, 1, 2, 3, and 4 cans. Each level of cans was tested 3 times, so there were 3 repeated trials. Again,

To find the **mean** for each level of the independent variable, find the sum of the measurements and divide that sum by the number of trials.

you could summarize the data by finding the mean travel time. Remember, mean is one way to determine the 'average.' For each level of the independent variable, find the sum of the measurements and divide that sum by the number of trials. Use this formula to check the means in Table 2.4.

You've probably heard the saying "You're all thumbs!" It means you're clumsy. But what would be the effect of "being all fingers" (and no thumbs)?! You'll need to use what you've learned about independent and dependent variables, constants, controls, repeated trials, hypothesis, and types of data to do *Investigation 2.2, You're All Fingers*. Read it over. Then state a hypothesis about what you think will happen.

Investigation 2.2

You're All Fingers!

WHAT YOU NEED

- Masking Tape
- Large shirt/jacket with 3 or more buttons
- Clock/watch with a second hand
- Partner to assist you

See Chapter 8, Experimenting Safely, Section E, Animals and Humans.

WHAT YOU DO

1. Put the shirt or jacket on over what you're wearing.
2. Measure the time it takes to button the buttons on the front of the shirt. Enter your data in the data table.
3. Have your partner tape your thumbs to the sides of the palms of your hand so that the thumbs won't move. Measure the time required to button the shirt or jacket. Record your data.
4. Repeat steps 1-3 for a total of 5 trials.

Method of Buttoning	Time (sec) Trials					Mean Time (sec)
	1	2	3	4	5	
Thumbs Free						
Thumbs Taped						

WHAT YOU LEARNED

1. What was your independent variable?
2. What was your dependent variable?
3. What were the constants?
4. How many repeated trials did you do?
5. Was your hypothesis supported by the data?
6. What was the control?
7. Which type of typical or central value, means or modes, would you use to summarize your data? Why?

Refining Experiments

For more information on Refining Experiments, use these search words or phrases:

"types of data"
"repeated trials"
"parts of an experiment"
"components of an experiment"

to find websites such as this:
hanover.k12.va.us/sjms/rhyne/backto.htm

Practice Problems

Refining Experiments

1. Identify the control in each of the following experiments. Tell if the control is a "no treatment control" or an "experimenter selected control."
   a. Shawnita investigated how well different kinds of soap would make oil and water mix. She put 240 mL of water and 5 mL salad oil in a zipper-lock bag. She shook the bag 30 times in 5 seconds. Then, she observed the number and size of the oil drops. Next, she repeated the experiment using 240 mL of water, 5 mL of salad oil, and 1 drop of dishwashing liquid. She also tested laundry detergent, shampoo, and liquid hand soap.
   b. Marques became interested in objects floating in different liquids. He put vinegar in a jar and measured how high a pencil would float in the jar. He repeated the experiment using other kinds of liquids: corn oil, motor oil, water, soda, and milk.
   c. Carlos became interested with the floating of objects in salty water. He put warm water in an olive jar and added 45 mL of iodized table salt. Then, he measured how high a pencil would float in the jar. He repeated the experiment using other kinds of salt—non-iodized table salt, Kosher salt, ice cream salt, and salt to de-ice steps.

2. Identify the kind of data collected in the following experiments. Indicate if the data are measurements, observations, or counts.
   a. Maria held her arms straight in front of her and wiggled her fingers. She kept looking straight ahead. She moved her arms to the side until she could no longer see her wiggling fingers. She had her sister measure the distance between her hands in centimeters. Maria repeated the experiment with her right eye covered and then with her left eye covered.
   b. Wade performed a taste test to determine which brand of cola his family preferred. He blindfolded each member and had them taste 5 brands—A, B, C, D, or E. Each person described the taste of each brand of cola as very good, good, not so good, or bad.
   c. Jerann investigated the effect of microwaving radish seeds on their growth. She exposed 4 different groups of seeds to 15, 30, 45, and 60 seconds of microwave radiation. She used a setting of 3 on the microwave. Seeds were placed in the same type of container and in the same location in the microwave. Jerann planted the seeds. At the end of 2 weeks, she counted the number of seeds that germinated. She also measured the height in centimeters of the plants.

3. Suggest types of data that you could collect in the following experiments:
   a. the effect of different sounds on dogs;
   b. the effect of various temperatures on soda pop;
   c. response of the human eye to different lights.

4. Classify the data in the following experiments as measurements, counts, or observations:
   a. the effect of different strengths of bleach on the time for stains to disappear;
   b. the ability of various metals to conduct/not conduct electricity;
   c. the number of roots produced with different concentrations of a rooting hormone;
   d. the amount of water (mL) consumed by chickens from red, blue, and yellow bowls;
   e. the influence of phosphorous on the color of algae in an aquarium.

5. For each of the experiments in Question 4, indicate the most appropriate way to summarize repeated trials: a) calculate the mean; b) identify the mode.

   For Experiments 6 and 7:
   a. identify the independent variable, dependent variable, control, constants, and number of repeated trials;
   b. classify the data collected as measurements, counts, or observations;
   c. indicate how you would summarize the data: calculate the mean, or identify the mode.

Remember

The **mean** is the sum of the measurements divided by the number of trials.

The **mode** is the most frequent observation.

6. Lillian explored the magnifying power of water drops. She bent the end of a paper clip into a 5 mm wide circle. She filled the circle with water and looked through the drop at the letter E. She tried 5 different drops; each time she noted the magnification as none, small, or large. She repeated the experiment using a 10 mm circle with water, and a 15 mm circle with water.

7. Marlo investigated the effect of ear lengths on the ability to hear. First, he had a friend walk toward him with a ticking clock. He measured the distance in meters (m) at which he first heard the clock. Then, he made 3 sets of ears—a 12 cm pair, a 24 cm pair, and a 36 cm pair. He repeated the experiment using each set of ears.

3

Analyzing Experiments

Diagrams, Flaws, and Improvements

At this point you know the parts of an experiment and what they mean. Suppose a scientist did an experiment that involved, 'amount of water.' Then she asked you, "Was the amount of water the independent variable, the dependent variable, a constant, or the control in my experiment?" What would you say? It's a tough question. In fact it is impossible to answer that question given what you were told. The amount of water could be any of the choices depending on its role in the experiment. An experimental design diagram will help you to identify each component of an experiment by determining its role. Best of all, an experimental design diagram does not take much time to construct.

Speaking of time, have you ever watched a grandfather clock, the kind of clock with a long pendulum that swings back and forth ever so slowly and smoothly? Boring.... OK, the pendulum's movement is boring. So let's design an experiment to see if we can make a pendulum swing faster. We could change at least three things; the length of the pendulum, the mass of the swinging object, and the distance to the side where you release the pendulum. In *Investigation 3.1 Swinging!* we changed only the length of the pendulum. Read the experiment and then we'll make an experimental design diagram that will show the role of each component of the experiment.

TABLE 3.1 Checking the Experimental Design

Checklist Question Number	Questions
✓1.	Does the title clearly identify both the independent variable and the dependent variable?
✓2.	Does the hypothesis clearly state how you think changing the independent variable will affect the dependent variable?
✓3.	Is there just one independent variable? Is it well defined?
✓4.	Are the levels of the independent variable clearly stated? Are there enough levels of the independent variable tested? Are there too many?
✓5.	Is there a control? Is it clearly stated?
✓6.	Are there repeated trials? Are there enough of them?
✓7.	Is the dependent variable and how it is to be measured or described clearly identified and stated?
✓8.	Are the constants clearly identified and described? Are there any others?

Using the checklist, we identified the following problems and suggestions for improvement in the *Huff, Puff, and Slide* investigation:

✓1. Does the title clearly identify both the independent variable and the dependent variable?

Improvements: Change title to "The Effect of the Number of Pennies in a Container on its Ability to Slide." Huff, Puff, and Slide tells you nothing about the IV and DV.

✓2. Does the hypothesis clearly state how you think changing the independent variable will affect the dependent variable?

Improvements: None: hypothesis is O.K.

✓3. Is there just one independent variable? Is it well defined?

Improvements: The independent variable, number of pennies (mass), is O.K. The unit of mass in this experiment is the mass of a U.S. penny which is approximately the same for all pennies. To be more precise, you could measure the mass of the pennies using units such as grams (g).

✓4. Are the levels of the independent variable clearly stated? Are there enough levels of the independent variable tested? Are there too many?

Improvements: The levels are clearly stated, but 2-4-6-10 is missing levels 0 and 8 coins. The levels are usually set at equal intervals or multiples: 0, 2, 4, 6 or 0, 2, 4, 8, 16.

✓5. Is there a control? Is it clearly stated?

Improvements: The zero (0) coins level of the independent variable—no pennies—should be added and labeled as the control.

✓6. Are there repeated trials? Are there enough of them?

Improvements: There's a problem here! Only one trial was done. When you conduct repeated trials, you test each level of the independent variable several times. A major way to improve this experiment would be to add another step to this procedure. Repeat Steps 1-4 four more times for a total of 5 trials.

✓7. Is the dependent variable and how it is to be measured or described clearly identified and stated?

Improvements: The directions say measure the distance the container slides. Specify the units you will use to measure, in centimeters or in inches. Also tell which path you will measure—a straight line or the actual path the container took.

✓8. Are the constants clearly identified and described? Are there any others?

Improvements: The constants could be more clearly stated. Define the smooth surface, for example, a polished wood dining table or Formica counter top. Define the type of container, such as 8 oz. plastic cup. Tell where to aim the blow—at base or mid-container. Tell if the container was covered or uncovered.

So what would the new and improved version of *Huff, Puff, and Slide* look like if you wrote it knowing what you do now? See the revised experimental design diagram on the next page.

Analyzing Experiments **31**

Title: The Effect of the Number of Pennies in a Container On Its Ability to Slide

Hypothesis: If the number of pennies of a container is increased, then the distance it will slide will decrease.

IV: Number of Pennies				← Independent Variable
0 Pennies	2 Pennies	4 Pennies	6 Pennies	← Levels of independent variables including the control
5 Trial	5 Trial	5 Trial	5 Trial	← Repeated Trials

DV: Distance container slides (cm) ← Dependent Variable

C: 250 mL (8 oz.) plastic cup
Blow at base of container ← Constants
Polished wood tabletop
Same person blowing

Activity 3.1 Improving Investigations

For each of the following two investigations:

A. construct an experimental design diagram using the directions in Figure 3.2;
B. use the checklist in Table 3.1 to determine ways to improve the experiment; and
C. check your work against the suggested answers.

Investigation 1: Zöe was looking through a cardboard cylinder of a roll of paper towels. She thought that it restricted her angle of vision and wondered what the effect of long and short hand-held tubes would be. She made tubes of equal diameter that were 5, 10, 15, and 20 mm long. She then stood 10 m from a wall, held the tube up to her eye, and tried to determine how much of the wall she could see. She measured the width of the section of the wall that she could see.

Investigation 2: Zeke surprised his mother by voluntarily doing the dishes! Unfortunately, he used laundry detergent, and soap suds gushed everywhere. That is when Zeke learned that different cleaning products produce very different amounts of suds. He studied the properties of laundry detergent, shampoo, powdered hand soap, and liquid hand soap. Zeke put 500 mL of distilled water into 4 empty, clear, 1 liter soda bottles. He added a different kind of cleaning detergent to each, shook the bottle for 30 sec, and then measured the height of the suds.

Investigation 1

Zöe was looking through a cardboard cylinder of a roll of paper towels. She thought that it restricted her angle of vision and wondered what the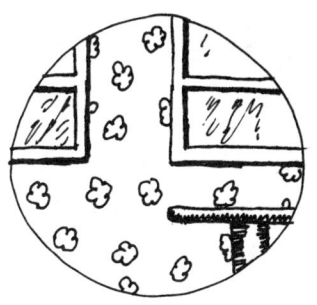
effect of long and short hand-held tubes would be. She made tubes of equal diameter that were 5, 10, 15, and 20 cm. long. She then stood 10 m from a wall, held the tube up to her eye, and tried to determine how much of the wall she could see. She measured the width of the section of the wall that she could see.

Title: The Effect of the Length of a Cylinder on the Width of a Field of Vision

Hypothesis: If the length of a cylinder is increased, then the width of the field of vision is decreased.

IV: Length of a Cylindrical Tube (cm)			
5 cm	10 cm	15 cm	20 cm
1 Trial	1 Trial	1 Trial	1 Trial

DV: Width of field of vision (cm)

C: Distance to wall
Diameter of tube
Hand-held tube

Checklist

Number	Title	Improvements
✓5.	Control	Control needed—no tube
✓6.	Repeated Trials	More repeated trials needed—5
✓8.	Constants	Fixed platform holder for tube rather than hand-held

Analyzing Experiments

Investigation 2

Zeke surprised his mother by voluntarily doing the dishes! Unfortunately, he used laundry detergent, and soap suds gushed everywhere. That is when Zeke learned that different cleaning products produce very different amounts of suds. He studied the properties of laundry detergent, powdered hand soap, and liquid hand soap. Zeke put 500 mL of distilled water into 4 empty, clear, 1 liter soda bottles. He added a different kind of cleaning detergent to each, shook the bottle for 30 sec, and then measured the height of the suds.

Title: The Effect of the Type of Detergent on the Amount of Suds Produced
Hypothesis: If the type of detergent is varied, the order of sudsing will be laundry detergent, shampoo, powdered hand soap, liquid hand soap.

IV: Type of Cleaning Product			
Laundry Detergent	Shampoo	Powdered Hand Soap	Liquid Hand Soap
1 Trial	1 Trial	1 Trial	1 Trial

DV: Width of field of vision (cm)
C: Distance to wall
Diameter of tube
Hand-held tube

Checklist

Number	Title	Improvements
✓3.	Independent Variable	Independent Variable—Be specific about brands and type, for example, Laundry Detergent should be Tide Powder Concentrate.
✓4.	Levels of Independent Variable	Equal amount of detergents must be used, such as, 15 mL each.

✓5.	Control	Control needed such as distilled water with no amount of soap.
✓6.	Repeated Trials	Need 5 or more repeated trials.
✓7.	Dependent Variable	Units for dependent variable needed.
✓8.	Constants	Shake rate better defined—5 shakes/sec for 15 sec.

For more information on Analyzing Experiments, use these search words or phrases:

"experimental design diagram"
"parts of an experiment"
"components of an experiment"

to find websites such as this:
fayar.net/east/teacher.web/Math/young/DwD/Workshops/Leadership%202001/EXPERIMENTAL%20DESIGN%20OVERVIEW.doc

Practice Problems

Diagrams, Flaws, and Improvements

For each of the following examples:

A. construct an experimental design diagram;

B. use the checklist to determine ways to improve the experiment; and

C. check your work against the suggested answers provided by your teacher.

1. Zelda decided to see if some breeds of dogs learn behaviors faster than others. She used her St. Bernard, and asked her neighbors if she could borrow a Poodle, a German Shepherd, a Chihuahua, and a Fox Terrier for an hour each. In the basement of her home, she taught them to sit and then to shake hands, using "Dawg Treats" as a reward. She repeated her "lessons" to each dog 2 times.

2. Dion noticed that his trophies often needed to be dusted. He wondered if changing the shelf height would affect the amount of dust on a shelf. He collected 14 coat hangers. He cut 14 identical pieces of wax paper, covered them all lightly with Vaseline, and attached the paper to the coat hangers. He hung 2 hangers so that the bottom of the paper was at 0, .3, .6, .9, 1.2, 1.5, and 1.8 m off the floor in the hallway. A week later, he took the hangers down. Holding the greased wax paper in front of a bright light, he compared amounts of dust collected at each height.

36 Chapter 3

Finding Project Ideas

Interests, Labs, Books, and Magazines

Scientists get ideas for experiments in many ways. They use their senses to observe the world and they read about science too. Scientists do research in both the library and the laboratory and they also attend meetings where they share research ideas. Scientists study reports of experiments by others and sometimes get ideas for investigations in the unanswered questions in someone else's work.

You too can get ideas for experiments in many ways. Are there things you wonder about? Observe the world around you or visit a library. Libraries are filled with books and magazines on science. The books of science activities, tricks, and magic will amaze you and you will find fascinating stories in science magazines. Try reading a little about things that interest you. Think about how you use your free time. Are there experiments hidden in your hobbies?

Using Your Interests

Being interested in the things you do makes them fun. A science project that interests you can be fun, too. It may not be easy, but something that is too easy is no fun. You will enjoy your project if you are interested in the topic you choose. Before choosing a topic to investigate, think about what already interests you. What subjects do you most enjoy? What do you like to do in your free time? What are your hobbies? What sports do you like to play? How about special talents like art or music?

Starting with something that interests you is a good place to begin. What have you wondered about that is related to an interest of yours? For example, if you were interested in running, have you ever wondered why runners begin some races standing up and others in a crouched down position? Does the position make any difference? What would happen if runners always started standing up?

Not many of us like doing household chores. If you would like to find ways to make them easier, studies of products may interest you. Do some brands of paper towels clean better? Does stainguard in dryer-static sheets really work? Are there ways you can finish a chore more quickly? Think about mowing the lawn. What affects the time to mow the lawn? Look at your chores for science project ideas.

You can also use hobbies and clubs as sources of ideas for experiments. If you play with Frisbees, think about what affects their flight. Do other brands fly the same? How does the angle of throw affect the flight? If you are a scout, is there a merit badge activity you can modify into an experiment? For example, you could investigate ways to collect solar energy, while earning your outdoor cooking badge.

Books of Science Activities, Tricks, and Magic

There are many books in science that are collections of science activities, tricks, and magic. Books like these can be found in the science section of libraries. Although most of the activities in these books are presented as demonstrations, the activities can easily be changed into interesting experiments.

Did you know that you can get electricity from a lemon? Read the science activity given in ***Activity 4.1, An Electric Lemon***. As you read, note the variables involved. Ask yourself, "What could I change in this investigation that might affect the results?" Some possible questions are listed below.

> Do different kinds of lemons act differently?
>
> How about other citrus fruit?
>
> Does the temperature of the fruit make a difference?
>
> What about the kind of metal objects used?
>
> Could you use a small light bulb or a voltmeter to measure the response?

FIGURE 4.1 INTEREST INVENTORY

1. What is your favorite hobby?

2. What sports interest you?

3. What is your favorite school subject?

    Science
    Mathematics
    Art or music
    English or social studies
    Health and physical education
    Technology

4. If science is one of your favorite subjects, which science do you like best?

Biology:	the study of living things—plants, animals, bacteria, fungi, protozoans
Chemistry:	the study of how materials are formed and changed—dissolving, evaporating, burning, tarnishing, digesting
Physics:	the study of energy and its changes—light, electricity, machines
Earth Science:	the study of the earth and its place in the universe—weathering, erosion, humidity, solar energy
Environmental Science:	the study of interactions of living things and their environments—forest, pesticides, pollution

5. Where would you like to work on your project—home or school?

    home
    school

6. What careers have you thought about?

7. What interesting books have you read on a science topic?

8. What films or television programs have interested you?

9. To what clubs or organizations do you belong?

© Cothron, Giese, & Rezba. *Teacher's Guide: Science Experiments by the Hundreds.* Kendall/Hunt. 1996

Finding Project Ideas

Activity 4.1 A Typical Activity Found in Library Science Books

An Electric Lemon

Materials:
1 lemon
1 galvanized nail
1 piece of bare copper wire
1 radio earphone plug

Instructions:
1. Make the lemon juicy inside by rolling it hard against a surface.
2. Carefully cut two slits in the lemon along one side of the lemon, about 3 cm apart.
3. Put the nail in one slit and the bare copper wire in the other slit. Be sure the nail and wire do not touch inside the lemon.
4. Touch the earphone plug to both the nail and the wire at same time. You should be able to hear static sounds in the earphone indicating that an electric current is flowing.

Reason: The lemon juice reacts differently with the two metals causing a small voltage difference between the metals.

You may find it helpful to think about causes and effects when reading activities. In the lemon experiment, the cause is the interaction of unlike metals and lemon juice. The result is a different voltage produced by two different metals. Think of causes as independent variables and effects as dependent variables. Then ask yourself some questions: "What are some ways I could change the causes?" or "What are some other ways I could measure the effects?" Thinking of how to change the causes and how to measure the effects may get you just the ideas you need for an experiment. Before you experiment with chemicals or electricity, read Section A, Chemicals and Section B, Electricity in Chapter 8, Experimenting Safely.

Science Textbooks and Laboratory Activities

Science textbooks and laboratory manuals are good sources of ideas for science projects. Some students get an idea for a science project from reading a chapter. Other students get ideas by wondering what would happen if they changed the variables in a lab activity.

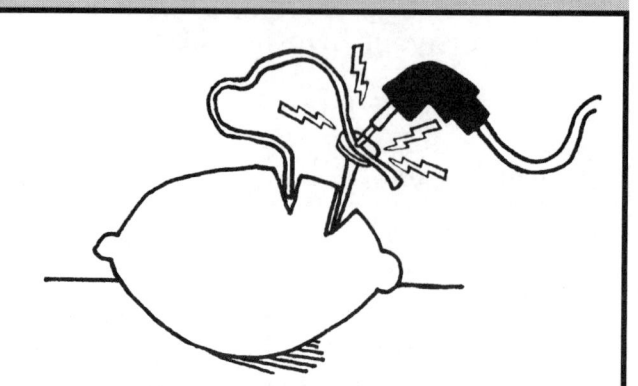

Textbooks are not usually found in school libraries. Ask your science teacher about textbooks you could borrow. Look through books from both higher and lower grades; last year's science textbook is a good place to start. You know more about those science topics than the science you will be studying this year.

Activity 4.2, Growing Molds is a typical life science lab activity on molds. Read the directions for the lab and identify the variables. What variable is changed on purpose? What variable responds? What things are held constant? What could you change in this activity that might cause different results? Think about your own experiences with mold—probably at home in the bathroom or kitchen. Could you affect the growth of mold in your home? Before you experiment with molds, read about safety in Chapter 8.

Reading about Science in Magazines and Books

Libraries have books on almost every topic you can imagine. Some of these books are science activity books, but most of them are informational books on such topics as digestion, bees, magnetism, and electricity.

ACTIVITY 4.2 — A Typical Life Science Activity Found in a Textbook

Growing Molds

Purpose:
To determine what materials will grow molds.

Materials:
Items such as cheese, bread, potato, oranges, and jelly
Eye droppers
Disposable jars with lids
Hand lens

Instructions:
1. Place each of the items in a small jar. If the item is solid, add a few drops of water. Expose the items to the air for 48 hours, then seal each jar with a lid. Label each jar with the date and the item name.
2. Observe the sealed jars each day for one week and record changes in the items' appearance.
3. At the end of the week, use the magnifying lens to examine the sealed jar to find evidence of any mold growth. Draw what you see.

If a topic like digestion or magnetism interests you, use the card catalog or a computer search system to find books on that topic. When you locate books on your topic, skim them to see which ones you can easily read and understand. Look at the table of contents of each book and choose a chapter. Skim that chapter by reading the introduction and the summary. Turn the chapter headings and subheadings into questions. Look at the pictures and read the captions under them. Doing this may help you identify areas of interest to learn more about.

Science magazines like *National Geographic World*, *Ranger Rick*, *Science World*, *Kids Discover*, and *3-2-1 Contact* are found in most school libraries and are good sources of ideas. Reading an article about bloodhounds tracking people might give you an idea for an experiment. If dogs can recognize the scent of people, can people recognize the scent of particular dogs? You might design an experiment to see if people could recognize the scent of their own pets.

Using the Internet

The Internet is a wonderful tool for finding information. On the Web you can easily find numerous references on most any topic. A search, however, can often lead to an overwhelming number of sites. Having two skills will help you receive a manageable amount of information that meets your needs:

1. selecting appropriate search engines,

2. using the search features to quickly locate specific information and resources.

   Some popular search engines used by educators are Alta Vista, Northern Light, and Google. A useful site to help you locate and evaluate different search engines is Beaucoup (http://www.beaucoup.com). This site lists more than 2,000 search engines, indices, and directories, and organizes them into categories. In addition it provides reviews and tips for using many of the most popular search engines.

   Each search engine has its own set of rules for focusing a search that are explained through the 'help' or 'advanced search' buttons. Knowing how to use these advanced search options is what allows the Internet to be a very valuable tool. See also Chapter 6, Exploring a Project Idea, for additional suggestions on how to use the Internet to gather the information you want and need.

Summary

Sources of ideas are all around you. Independent variables that you can manipulate are everywhere. Observe the world, read science books and magazines, and talk with your parents and teachers about your ideas. Think about your interests, hobbies, or daily activities for potential ideas. Remember, it's important to pick a topic that interests you. Being interested in things makes them more enjoyable. Doing a science project that interests you will make it fun as well.

Using some of the idea sources described in this chapter may help you identify an area of interest or topic you want to explore. In Chapter 6, you will learn how to use the library and the Internet to learn more about an interesting topic. You can use what you learn to change a topic into an experiment.

Searching the Web

For more information on Finding Project Ideas, use these search words or phrases:

"science project ideas"
"science fair ideas"
"science experiments"
"elementary science projects"
"children's science projects"
"ideas for science projects"

or use specific topic experiments, such as:

"magnet experiments"
"consumer product experiments"

to find websites such as this:
ipl.org/div/kidspace/projectguide/

Practice Problems

Finding Project Ideas

1. **Project ideas from a hobby**: Can you help two roller skaters with some ideas for a project? Read the following and list at least three potential independent variables they could study. If you skate, include ideas you have from your own experience with skating.

   Matt and Katie both like to skate. They would like to do a science project together on their hobby. They each have collected several pairs of skates. Caring for these skates, like oiling the wheels, is important to them. They also spend time attending skating exhibitions and watching skating events on TV. Both of them have learned a lot about the different body positions men and women professional skaters use to start, turn, and stop.

2. **Project ideas from a textbook laboratory activity**: Read about speed in *Activity 4.3, Measuring Speed*. What could you change in this activity that might affect the speed? Think of as many ways as you can.

3. **Project ideas from a science activity library book**: Study the combined science and art activity in *Activity 4.4, Rising Colors*. What are some ways you could change this activity to get different results?

ACTIVITY 4.3   A Typical Life Science Activity Found in a Textbook

Measuring Speed

Purpose:
To determine the average speed of an object between two points.

Materials:
Ruler with a groove along its length
Marble
Stop watch
Meter stick
Books to lift ruler
Tape

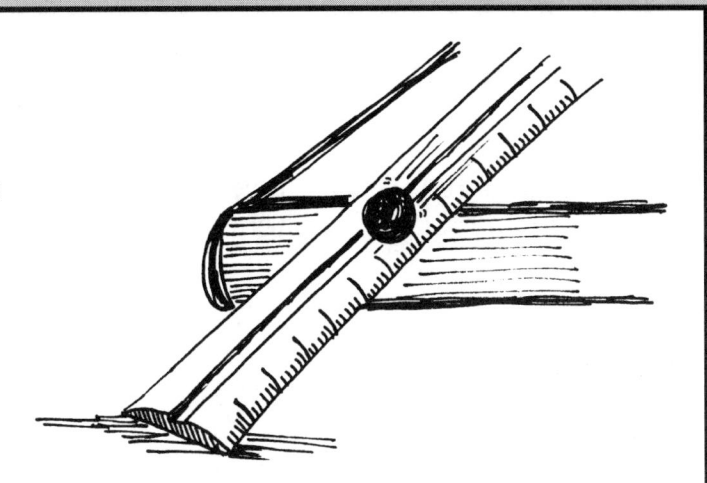

Instructions:
1. Make a ramp with the ruler, using a book as shown in the diagram.
2. Use the meter stick and the tape to mark a distance of 2 m. from the end of the ruler.
3. Place the marble at the top of the ramp/ruler and release it.
4. Time the seconds it took the marble to reach the 2 m point.
5. Repeat Steps 3 and 4 four more times for a total of 5 trials.
6. Calculate the speed of the marble for each trial by dividing the distance traveled by the time in seconds and record in a data table.

ACTIVITY 4.3 A Typical Life Science Activity Found in a Textbook

Measuring Speed

Materials:
Colored markers (water soluble)
Lab filter paper or coffee filters
Clear glass
Water

Instructions:
1. Cut the filter paper into 3 cm strips.
2. Place a strip in a glass so the end of the strip rests on the bottom. Bend the top of the strip over the rim of the glass.
3. Remove the strip and make a dot of ink 5 cm from the bottom.
4. Put 3 cm of water in the glass and place the strip in the glass.
5. Observe how the rising water acts on the color dot.
6. Repeat the activity with other colored inks.

Reason:
The colors in inks and dyes are often combinations of coloring substances. When dissolved in water, these colors can be separated because some color molecules are bigger and heavier than others. Some move faster up the paper strip, while others move slower.

Brainstorming Project Ideas

Four Question Strategy

Heads I win, tails you lose. Examine the two sides of a penny: look at the texture, design, and indentations. On which side of a penny can you heap more drops of water? Try **Investigation 5.1, Drops on a Coin**.

Investigation 5.1

Drops on a Coin

WHAT YOU NEED

- Penny
- Water
- Paper towel
- Metric ruler
- Pipette or medicine dropper
- Safety goggles

See Chapter 8, Experimenting Safely, Section A, Chemicals.

WHAT YOU DO

1. Place a penny on the paper towel with the heads-side up.
2. Fill the dropping device with water and hold it about 3 cm above the coin.
3. Count the number of drops added until the water spills over the side of the penny.
4. Record the number of drops in Table 5.1.
5. Dry the penny and repeat 2 more times for a total of 3 trials.
6. Repeat steps 1-5 using the tails-side of the penny. Record your data. Calculate the mean number of drops you placed on each side.

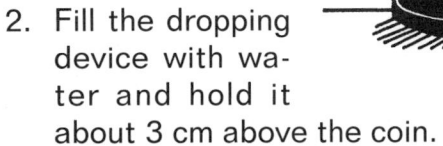

$$\text{Mean} = \frac{\text{Trial 1} + \text{Trial 2} + \text{Trial 3}}{3}$$

TABLE 5.1 The Effect of the Side of a Coin on the Number of Water Drops That Can Be Placed on It

Side of Penny	Number of Drops Trials			Mean Number of Drops
	1	2	3	
Heads				
Tails				

Were your results what you expected? Would you get the same results with salt water, soapy water, quarters, or different droppers? What else might affect the number of drops?

Asking questions about the penny activity probably gave you many ideas for experiments. To design a good experiment, however, you need to focus your ideas. It isn't hard. In fact, it's as easy as answering four questions. Here's how the Four Question Strategy works.

Four Question Strategy

Say you're interested in designing an original experiment with liquid drops on coins. The first question to ask yourself is:

1. **What materials are readily available for conducting experiments on (liquid drops on coins)?**

    Possible answers:
    Coins
    Liquids
    Droppers
    People
    Cleaners

The more things you list, the better the experiment you will be able to design. Try to pick materials that are inexpensive and easy to find. You may be able to borrow materials from your school, parents, or people in the community.

Next, ask yourself:

2. **How do (liquid drops on coins) act?**

   Possible answers:
   Liquids heap.
   Liquids spill.
   Liquids make wet circles on paper.

The idea is to brainstorm as many ideas as you can. This is a time to think of all the possibilities. After you have finished brainstorming as many actions as possible, stop. Select the one action upon which you will focus. Continue brainstorming by asking yourself the next question:

3. **How can you change the set of (liquid drops on coins) materials to affect the action?**

   Possible answers:

Coins	**Liquids**	**Droppers**	**People**	**Cleaners**
Kind	Kind	Kind	Gender	Type
Age	Temperature	Size of opening	Hand used	Brand
Cleanliness	Source	Height above coin	Manual skills	Time applied
Sides	Additives			Method of application
Metal	Color			

What you now have are lists of potential variables or things you could change. The longer the lists, the more choices you'll have. Each variable you generate is a possible independent variable. Select one variable you want to manipulate (say, the kind of liquid). It will be your independent variable; all the rest must become constants in your experiment. Finally, ask yourself:

4. **How can you measure or describe the response of (liquid drops on coins) to the change?**

   Possible answers:
   Count the number of drops.
   Describe the shape of the liquid on the coin.
   Measure the height of the heap.
   Measure the diameter of the liquid spill on the paper.

This final question helps you decide how to measure or describe changes in liquids placed on coins. Choose one of these ways to measure or describe the change in the action you selected in Question 2 (say, the diameter of the liquid spill in centimeters). The quantity or quality you select is the way you will describe or measure your dependent variable.

So, to begin a science project, all you have to do is pick one independent variable from Question 3 (such as, kind of coin) and one dependent variable from Question 4 (how about—count the number of drops). Remember what happens to all the other variables from Question 3. You must keep them the same in your experiment; they are your constants.

Skateboards (Ideas from Hobbies)

The Four Question Strategy helped you think of many ways to experiment with coins. But, would the strategy work with other topics as well? What if you wanted to experiment on skateboards? Let's try it.

1. **What materials are readily available for conducting experiments on (skateboards)?**

    Possible Answers:
    Skateboards
    Wheels
    Ramps

2. **How do (skateboards) act?**

    Possible Answers:
    Skateboards travel long distances.
    Skateboards go fast.
    Skateboards turn and flip.

3. **How can you change the set of (skateboard) materials to affect the action?**

    Possible Answers:

Skateboards	Wheels	Ramps
Cost	Diameter	Height
Brand	Width	Length
Material	Tread design	Curvature
Length	Material	Surface
Width	Age	

4. **How can you measure or describe the response of (skateboards) to the change?**

   Possible Answers:
   Measure the total distance traveled.
   Measure the time to go a certain distance.
   Describe the ease in changing direction.
   Measure the distance it will stay airborne off a ramp.
   Measure how long the parts last.

Selecting a project to do on skateboards is as easy as:

◆ choosing one variable from Question 3 (say, wheel diameter) as your independent variable;

◆ choosing a variable from Question 4 (how about, distance) as your dependent variable;

◆ keeping all other variables in Question 3 the same; they become your constants.

As you search for a topic of interest, you may find a hobby or a favorite activity like skateboarding that interests you and that you can turn into a project. The Four Question Strategy worked for skateboards. Let's try it for plants. Almost everyone grows plants at home. Suppose you wanted to use plants in your experiment. What could you do?

Plants (Ideas from Home)

1. **What materials are readily available for conducting experiments on (plants)?**

   Possible Answers:
   Soil   Light
   Water   Fertilizer
   Seeds   Containers

Brainstorming Project Ideas

2. How do (plants) act?

   Possible Answers:
   Plants grow.
   Plants bloom.
   Plants produce fruit.
   Plants die.

3. How can you change the set of (plant) materials to affect the action?

   Possible Answers:

Water	**Seeds**	**Containers**
Amount	Kind	Materials
Composition	Depth	Number of holes
Scheduling	Spacing	Size
Source	Age	Shape
Method of application	Size	Location of holes

Soil	**Light**	**Fertilizer**
Kind	Intensity	Amount
Amount	Source	Composition
Compaction	Color	Kind
	Schedule	Schedule
		Method of application

4. How can you measure or describe the response of (plants) to the change?

   Possible Answers:
   Measure the height of the stem.
   Count the number of flowers.
   Count the number of leaves.
   Determine the color of the leaves.
   Measure the diameter of the stem.

Fishing (Ideas from Television and Newspapers)

You can also use the Four Question Strategy to turn an idea from a television program or from a newspaper article into an experiment. Consider the following newspaper article:

> Wharf City: Fishing guide, Captain Suzi Sinker, reported that fishing season is in full swing. When asked the best way to fish, she replied, "Time is important; the hour just after sunrise and just before sunset are best." She also said, "use blue colored artificial bait or six inch blue plastic worms. Cast your lures to within a foot of the shore."

Let's use the Four Question Strategy on this article.

1. **What materials are readily available for conducting experiments on (fish)?**

   Possible Answers:
   Lures
   Time
   Casting techniques

2. **How do (fish) act?**

   Possible Answers:
   Fish bite.
   Fish ignore bait.
   Fish throw lures.

3. **How can you change the set of (fish) materials to affect the action?**

   Possible Answers:

Lures	Time	Casting Techniques
Size	Length of time	Location
Shape	Time of day	Distance
Color		Arc/curve
Type		Hand motion

4. **How can you measure or describe the response of (fish) to the change?**

   Possible Answers:
   Count number of bites.
   Count number of fish caught.
   Measure the length of fish caught.

Brainstorming Project Ideas

Collapsing Bottles (Ideas from Books)

Suppose you read the following in a library book of science tricks, magic, and activities:

1. Place a small wad of very fine uncoated steel wool (such as used to refinish furniture) in a clean, clear, plastic soft drink bottle.

2. Add 60 mL of warm water, cap the container, and shake the container.

3. Check the container periodically for several hours.

4. The sides of the container will collapse.

To turn a demonstration into an experiment, use the same Four Question Strategy.

1. **What materials are readily available for conducting experiments on (collapsing bottles)?**

    Possible Answers:
    Plastic soft drink container
    Liquid
    Steel wool

2. **How do (collapsing bottles) act?**

    Possible Answer:
    Bottle collapses.

3. **How can you change the set of (collapsing bottle) materials to affect the action?**

    Possible Answers:

Plastic container	Liquid	Steel Wool
Size	Kind	Amount
Brand	Amount	Kind
Age	Temperature	Location in bottle
Color	Method of mixing	Coarseness

4. **How can you measure or describe the response of (collapsing bottles) to the change?**

   Possible Answers:
   Measure the time for the reaction to begin.
   Measure the total time for the bottle to collapse.
   Describe the amount of collapse.
   Measure the depth the bottle caved in.
   Describe the appearance of the contents of the bottle.

Practicing the Four Question Strategy

Topics to investigate are everywhere. Apply the Four Question Strategy to one of your interests, and presto, you are on your way.

For practice, try using the Four Question Strategy to brainstorm ideas for experiments on bubbles.

1. **What materials are readily available for conducting experiments on (bubbles)?**

   Possible Answers:

   _____
   _____
   _____
   _____

2. **How do (bubbles) act?**

   Possible Answers:

   _____
   _____
   _____
   _____

Brainstorming Project Ideas 55

3. **How can you change the set of (bubble) materials to affect the action?**

   Possible Answers:

   _____   _____   _____
   _____   _____   _____
   _____   _____   _____
   _____   _____   _____
   _____   _____   _____

4. **How can you measure or describe the response of (bubbles) to the change?**

   Possible Answers:

   _____
   _____
   _____
   _____

Compare your responses with other students in your class, or with the sample responses provided by your teacher.

Sometimes the toughest part of a science project is coming up with an experiment. The Four Question Strategy solves that problem, but creates another problem. The strategy produces hundreds of possible projects and you only have time to do one!

56   Chapter 5

Searching the Web

For more information on Brainstorming Project Ideas, use these search words or phrases:

"four question strategy"
"brainstorming project ideas"

to find websites such as this:
science-house.org/workshops/web/4question.html

Practice Problems

Brainstorming Project Ideas

1. Suppose you read the following article in your local newspaper. Use the Four Question Strategy to brainstorm ideas for experiments based on the article.

    Mr. I.B. Beat, a self-proclaimed exercise specialist, addressed the Pulse City Health Club. His topic was "Exercise and the Heart." He stated that:
    - The faster you walk or jog, the faster your heart will beat.
    - Using a step of 10 or 20 cm in height for step aerobics affects the heart rate the same.
    - The temperature and humidity of the exercise room do not affect heart rate for a given exercise.
    - Women's heart rates do not change as much as men's for a given exercise.
    - The longer you perform an exercise, the faster your heart will beat.
    - Situps and jumping jacks increase the heart rate by the same amount.

    In the question and answer period that followed, Dr. A. Orta, a heart specialist, challenged several of Mr. Beat's statements.

2. Brainstorm ideas on the following:
    a. home insulation materials;
    b. sodas (pop).

6

Exploring a Project Idea

Libraries, Note Cards, and Interviews

Oops! You've just knocked over a liter of your favorite soft drink and the spill is rapidly crossing the counter to your parent's freshly scrubbed floor. On television, the commercials make paper towels look great, but how about other readily available spill catchers such as napkins, newspapers, or paper bags? Let's test their effectiveness in *Investigation 6.1, Spill Catchers*.

Investigation 6.1

Spill Catchers

What You Need

- Various types of paper such as a paper towel, napkin, brown paper bag
- Scissors
- Container such as a glass, plastic cup or a beaker of water
- Metric ruler
- Pencil

SAFETY: Be careful using scissors. See Chapter 8, Section A, Chemicals.

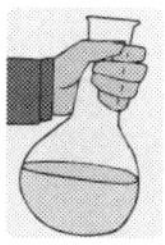

What You Do

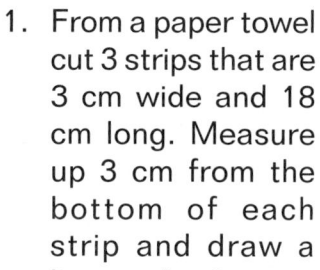

1. From a paper towel cut 3 strips that are 3 cm wide and 18 cm long. Measure up 3 cm from the bottom of each strip and draw a line at the 3 cm mark. Place the strip in water to that line.
2. At the end of 30 sec, quickly remove the strip from the water and mark the height the liquid rose. Measure the distance between the two marks in millimeters. Record your data in a data table.
3. If time allows, repeat Steps 1-2 four more times for a total of 5 trials and record your data. Or combine your group's data with other groups' in a class data table.
4. Repeat Steps 1-3 using paper from a napkin and brown paper bag.

59

Here's how our data looked. How does your data compare? Suppose you were assigned the task of designing experiments that you could conduct with paper. What might you do?

TABLE 6.1 The Effectiveness of Various Papers in Absorbing Water

Type of Paper	Height Water Rose (mm) Trials					Mean Height (mm)
	1	2	3	4	5	
Paper towel	47	50	43	46	48	47
Napkin	22	21	19	20	22	21
Brown bag	5	3	2	4	2	3

Before you decide on an experiment with paper, you should collect some additional information. From Investigation 6.1 you learned a little about how water travels in strips of paper, but there is much more to learn. Paper allows people to package things, share information, and keep records. There are over 7000 different kinds of paper. Each paper has special characteristics and behaves differently.

> **Remember**
> The **mean** is the sum of the measurements divided by the number of trials.

Learning as much as possible about this familiar item may help you improve your first idea for an experiment. Additional information about paper will help you design better experiments about paper. It might even give you a totally different idea for an experiment. You can learn more about paper or almost any other topic by using the resources in your science classroom, the library, and your community.

You will need certain skills to obtain and keep track of the information you collect. Keeping records of what you read and the information you learned are two of these skills. You will also need to locate information and summarize the information you found.

Referencing Information

It is important that you keep a record of the sources of information that you use. The titles and authors of what you read are examples of reference information you need to record. You will need this information for two reasons. First, you may need to return to a resource to re-read it or to find additional information. The second reason is to let other people know where you found the information you used. When writing papers, people often use the work and ideas of others. This is fine as long as credit is given to the author.

The kind of information you record about the resources you use depends on the type of material. The format that is used to record this information is called the **reference style**. The examples in this book are written using the style of the Modern Language Association. For most school projects, you should record your references using the style that is taught in your school's English classes. Sometimes, however, a particular reference style is required by a science fair or other competitive event. If you are going to enter your experiment in a competition, check with your teacher about special requirements. For more information on how to reference books, articles in magazines and journals, newspaper articles, encyclopedias, and information from the Internet see *Appendix A: Using Style Manuals*.

Taking Notes

You will find it easier to take and use your notes if you record the information the same way each time. Index cards, approximately 5 in. by 8 in. are recommended. The kind of information that should be recorded and its location on the card are illustrated in Figure 6.1. Here's what each part of the card means.

FIGURE 6.1 SAMPLE NOTE CARD

> **Topic:** Manufacturing Paper
>
> **Card Number:** 1
>
> **Location:** School Library
>
> **Call Number:** Reference Section
>
> **Reference Documentation:** Author: John B. Woods. Title: <u>Paper Manufacturing Processes</u>. Place: Raleigh, N.C. Publisher: Wood Research Institute. Copyright: 2003. Page Numbers: pp. 14-17.
>
> **Points of Interest**
> 1. Made from a plant carbohydrate called cellulose (p. 14)
> 2. Made from trees, cornstalks, rice and wheat straws, hemp, jute (p. 15)
> 3. Four manufacturing steps (pp. 16-17)
>   - Make wood pulp by removing tree bark, cutting logs into wood chips, breaking chips into fibers
>   - Improve wood pulp by screening unwanted materials, air-fluffing, and treating with chemicals
>   - Use screen belt, rollers, and heat to remove excess water from pulp
>   - Change pulp to paper by rolling, drying, pressing
>
> **Other References**
> Paper in the New Millennium. <u>The Sun City Times</u>. August 18, 2002.

- **Topic** is the major idea or subject described in the resource.

- **Card number** is the sequence number of the card for the same reference. If you have more than one card from the same reference, the number helps you keep them in order.

- **Location** is where the resource can be found such as the school or community library or your science classroom.

- **Call number** is the number used by the library to catalog and place the books on shelves; the number will help you find a resource if you need it again.

- **Reference documentation** is the information about the author, title, and so on that you will need for the reference style you have selected to use.

- **Points of interest** are the ideas you found important or interesting. Record the page number for each idea so that you can quickly come back to that idea later if you need more information.

- **Other references** are sources given by the author that interest you. Later, you may want to find the sources and read them.

Read the paragraphs from a book on molecular attractions in Figure 6.2. When you are through, look at the completed note card in Figure 6.3. See how information from these paragraphs was recorded on the note card. Notice what was selected for the points of interest. Were there other parts of the article that interested you?

FIGURE 6.2 READING ON MOLECULAR ATTRACTIONS

Molecules of matter are attracted. The attraction between like molecules is called cohesion. Because of cohesion, molecules of water remain together rather than moving apart. The attraction between unlike molecules is called adhesion. Because of adhesion, water wets paper, glue sticks substances, and bandages attach to your arm. Adhesion increases with the closeness of contact. For this reason, one must press down with a pencil to make a mark on paper; liquids stick better than solids; and fine particles adhere better than coarse. Because of molecular attraction, liquids rise in small openings or tubes; the smaller the opening, the higher the liquid rises. This movement is called capillary action. First, the liquid adheres to the sides of the tube. This causes the liquid level to be higher on the edges than in the center of the tube. Because the liquid molecules also cohere, the liquid in the center is pulled up. Gradually the liquid creeps up the sides of the tube and pulls the center liquid along. Through capillary action, water rises in papers and cloths, oil soaks a lamp wick, and water moves through the soil to plant roots.

Paragraph from page 150, Chapter 5, The Chemistry of Water in Everyday Chemistry *by A.L. Sheeran and D.L. Jones. Universal Scientific Publishing Company in Chicago, Illinois. 2004. Found in Valley High School Library.*

Use a note card to take notes on one of the sources included in Figure 6.4. These selections will help you learn about other properties of paper. When writing your reference information, record the information in the same location on your note card. If you need help, see Figures 6.1 or 6.3. For examples of how to document other sources such as newspapers and encyclopedias, see Appendix A.

When you record the most important information, write it in your own words. Do not just copy the author's words. If the author's wording is so important that you need to copy it, be sure to put quotation marks around the quoted words. Make the cards neatly so that you can read them several weeks or months later. Remem-

Exploring a Project Idea **63**

FIGURE 6.3 SAMPLE NOTE CARD FOR MOLECULAR ATTRACTION

> **Topic:** Attraction of Molecules
>
> **Card Number:** 1
>
> **Location:** Valley High School Library
>
> **Call Number:** (none given)
>
> **Reference Documentation:** Author: A.L. Sheeran and D.L. Jones. Title: <u>Everyday Chemistry</u>. Place: Chicago, Illinois. Publisher: Universal Scientific Publishing Company. Copyright: 2004. Page Number: 150.
>
> **Points of Interest**
> 1. Cohesion is the attraction between like molecules. Example—water.
> 2. Adhesion is the attraction between unlike molecules. Example—bandages and arm.
> 3. Capillary action is how things move up small openings, like water in paper. Adhesion makes the water stick to the paper. Cohesion holds the water together so that it keeps moving.
>
> **Other References** (none given)

ber to check the spelling of the information you recorded and to follow the capitalization and punctuation rules for the reference style you are using.

Locating Written Information

Finding parts of books and articles that give you information about a topic of interest is like a treasure hunt. You need clues to find the right locations. You can begin by looking at the table of contents and index of a science textbook. Use your textbook or ask your teacher for other similar textbooks. Always begin with the easiest book.

Suppose you needed more information about how paper absorbs water. You probably wouldn't find the words, "water absorption of paper," in either the table of contents or the index of a science textbook. But you can often find other words that may help you find parts of the book that will help you learn more about water absorption. For example, you might find the following words or phrases that might be helpful.

FIGURE 6.4 READINGS ON PAPER

Types of Paper. Manufacturers combine various wood pulps and sizing agents. Major pulps include mechanical, chemical, and rag. Sizing agents are added to make paper more resistant to liquids. Sizing agents are necessary for ink to dry on top of the paper rather than soaking into it. Writing papers include newsprint, book or magazine paper, and stationery. Newsprint is a blend of three parts mechanical and one part chemical pulp. Books and magazine papers are usually made of chemical pulp, rag pulp, or a mixture. Smooth surfaces for inks are produced by adding glue, starch, clay, or other fine minerals. Bond papers, containing a watermark, are used for business stationery and records. Absorbent papers and tissues are made from chemical and rag pulps. Examples of absorbent papers are facial tissue, paper towels, toilet paper, and filter paper. Tissue includes napkins, paper table cloths, and wrapping paper. Manufacturers of boxes, paper bags, and wrapping papers use large amounts of the kraft chemical pulp. Carton and container paper consists of kraft chemical pulp, waste paper, or straw pulp. Waste paper comes from old newspapers and cartons.

Article on Manufacture of Paper from Wood Daily Times Newspaper. *August 15, 2001. Page 6B.*

Stress in Objects. When objects are pulled, pushed, heated, or cooled stress results. Stress causes fibers in the material to lengthen or to shorten. If too much stress is applied, the object can be deformed or break. The strength of various materials is often compared by determining the force required to pull the material apart. Thomas Young invented a value, called Young's modulus, for comparing strengths. Scientific handbooks contain tables showing Young's modulus for various materials. Sir Robert Hooke experimented with springs and other stretchable objects, such as a flat piece of paper cut into a spiral. He learned that the more weight added, the further the object stretched. When too much weight was added and then removed, the stretchable material would not return to its original shape and size. The object had reached its limit of elasticity. Different materials have different limits of elasticity; exceeding the limit causes the object to break. Scientists, architects, and engineers use Young's modulus and the limit of elasticity to decide the best materials for constructing objects.

Information from Young's Modulus, a section in The Engineering and Physics Encyclopedia. *Lawrence Publishing Company. 2002. San Francisco, California. page 1007. Reference section of Hillsdale County Library.*

Exploring a Project Idea

Table of Contents
Properties of Matter (Chapter 2)
The Chemistry of Water (Chapter 4)
Forces (Chapter 7)
Chemistry & Manufacturing (Chapter 17)

Index
Stretching of materials (pp. 208-210)
Elastic substances (p. 200)
Movement of water (p. 87)
Capillary action (p. 90)

Use the table of contents of a book to find each of the chapters. Quickly read the headings and subheadings for topics of interest. If you find a good section, read it carefully and take notes using the procedures previously described. Likewise, scan the pages listed for a specific topic in the index, such as movement of water. If you find relevant information, make a note card.

Part of a librarian's job is to help people find the books and magazines they need. If you don't know where to start, you might begin by asking a librarian: "What reference books do you have on (*paper*)?" "How can I learn more about (*different types of paper*)?" If you're not doing a project on paper, substitute your topic of interest in the questions.

The *Reader's Guide to Periodical Literature* is your best single source for finding information in general (non-technical) magazines. It can quickly tell you what articles on a particular subject were published in most general magazines. If you know an author's name, it can also tell you what else the author has written. The *Abridged Reader's Guide*, for the most used journals, is often found in schools. For specialized science magazines, you would need to consult special indexes like *Biological and Agricultural Index* and the *General Science Index* that are found in college libraries. Today, many of these reference guides are available on computer, as well as in print form. Through many on-line and CD-ROM computer data bases you can access information quickly.

You can also conduct interviews with members of the community. People who might help can be found at universities, industries, and government agencies. Depending on your topic you might also contact zoos, nature centers, or museums. For tips on contacting people and asking questions, see *Appendix B: Conducting Interviews*.

Locating Digital Information

Searching for information on the Internet can be quite an adventure. Finding Web pages or Web sites can seem very easy or impossibly difficult. Searching the World Wide Web is different from searching a library because the Web is not indexed with a standard vocabulary like a library is. In a library you can use standardized Library of Congress subject headings to help you find information. When searching the Web, you have to guess what subject words will be on the Web pages you want to find. Here are some questions you might ask yourself to think of search words that could help you find information on your chosen topic, such as manufacturing paper:

1. What special words, names, and abbreviations are related to your topic?
   For example: cellulose, pulp,

2. Do you know of any groups, organizations, or museums that are related to your topic?
   For example: The Institute of Paper Science and Technology, The Robert C. Williams American Museum of Papermaking

3. Are there words that are likely to be in any Web document on your topic?
   For example: paper

4. Do any of the words commonly appear together?
   For example: "wood pulp," "making paper," and "manufacturing paper"

When using a particular search engine, learn their rules for focusing or improving a search. You can find the rules and other search suggestions by clicking on the 'help' or 'advanced search' buttons. An example of one of these rules is the use of quotation marks. Words in double quotes, for example "wood pulp," will appear together in all results exactly as you entered them. Other common rules to improve searches use symbols and words, such as +, –, ~, and, or. Each word or symbol changes how the search is made. Learning to use these words and symbols correctly will greatly improve your search for information on your topic.

Brainstorming Ideas

Now that you've become an expert on the subject of paper, it's time to use that new-found knowledge to brainstorm ideas for an experiment with paper. The more you know about a topic, the better the results will be.

Begin by returning to the Four Question Strategy described in Chapter 5. When responding to the four questions, use the new information you learned from the library. In the following example, we have shown library information in parentheses.

1. **What materials are readily available for conducting experiments on (paper)?**

    Possible answers:
    Paper
    Liquids
    Forces
    Inks
    Various writing instruments

2. **How does (paper) act?**

    Possible answers:
    Absorbs liquids
    Tears
    Absorbs/repels inks
    Stretches
    Provides surface for writing instrument

3. **How can I change the set of (paper) materials to affect the action?**

    Possible answers:

Paper	Liquids
Kind of pulp (*chemical, mechanical, rag*)	Kind
Amount of sizing	Temperature
Kind of sizing	Amount
General type of paper (*writing, absorbent, boxes, cartons*)	Closeness to paper
Type of writing paper (*newsprint, books, stationery, bond*)	
Type of absorbent paper (*facial, towels, toilet*)	
Type of tissues (*napkin, wrapping, table cloth*)	
Type of cellulose (*trees, rice, hemp*)	
Age	

Forces	**Ink**	**Writing Instrument**
Amount	Kind	Type of instrument
Direction applied	Amount	Brand
Dampness of paper when force applied	Method of applying	Sharpness of point
Kind of force	Fineness of ink particles	Age
		Temperature
		Pressure applied

4. How could I measure or describe the response of (paper) to the change?

Possible answers:
- Amount of liquid absorbed
- Time to absorb liquid
- Amount of force required to break (*stress*)
- Amount of stretch with certain force (*elasticity*)
- Time for ink to dry
- Amount of ink spread/absorption
- Adherence/non-adherence of particles (*cohesion, adhesion*)
- Maximum height liquid rose (*capillary action*)
- Clearness of words written on surface
- Time writing instrument will continue writing

By reading about a subject you can be more specific in your brainstorming. For example, you could subdivide paper into general categories—such as writing, absorbent, boxes, and cartons—and brainstorm variations within these large groups. Similarly, the list of ways that you can measure or describe the response can be expanded to include behaviors that you learned about in your readings such as capillary action, stress in objects, and elasticity.

Summary

Scientists get ideas for experiments in many ways. They use their senses to make observations of the world around them and they read about science. They learn as much as possible about a topic before deciding on an experiment. As they read, they record the sources of information used. In this chapter, you learned how to record information on note cards and how to locate information. You can learn about conducting interviews by reading Appendix B. With your expanded knowledge and ability to brainstorm, you are better able to explore a project idea of your own. Try your new skills with ***Investigation 6.2, Experimenting with Plants...Ideas, Ideas, and More Ideas***.

Investigation 6.2

Experimenting with Plants... Ideas, Ideas, and More Ideas

What You Need

- Index cards or paper
- Pen or pencil
- Readings in Table 6.4

What You Learned

1. How does your new brainstormed list of potential experiments on plants compare with the list in Chapter 5?
2. How did learning more about the topic change your ideas for experiments?

References

Gibaldi, Joseph. *MLA Handbook for Writers of Research Papers.* 5th ed. New York: MLA, 1999.

What You Do

1. Review the brainstormed responses about potential experiments on plants in Chapter 5.
2. Read the four readings on watering, African violets, fertilizer, and duckweed included in Table 6.4. Make a note card for each reading using the format that you've learned in this chapter.
3. Use the Four Question Strategy to brainstorm ideas for experiments with plants.

Remember

Refer to Figure 6.1 or page 62 for a sample note card.

FIGURE 6.5 READINGS ON PLANTS

Reading 1: Watering

More plants die from overwatering than underwatering. When the ground is too moist, the roots cannot get oxygen they need to survive. Once the roots begin dying, the leaves follow closely behind. Because the symptoms are similar, many people confuse overwatering with underwatering. Both conditions will cause wilting, stunted growth, and the leaves to turn brown, black, or yellow and drop off. With underwatering, the roots on the outside of the soil ball will be brown and shrunken; the stems will become soft and collapse near the bottom. Overwatered roots will be brown, mushy, and decayed. Unless plants have not been watered for a long time, the stems will not collapse.

When and how much to water plants depends upon the humidity, size of the pot, temperature, and amount of growth.

When the plant's environment is dry, warm, or, a crowded pot, more watering will be needed. Plants that are growing rapidly or blooming require substantially more water. As soon as the growth stops or the blooms die, the water level can be reduced.

In watering plants, the soil should be moistened all the way to the bottom of the pot. Lukewarm unsoftened water is recommended. Plants should be watered thoroughly each week rather than a small amount each day. When the plant is thoroughly watered, the water should drain through the hole in the bottom of the pot and accumulate in the saucer or ceramic overpot. To prevent overwatering, be sure to pour the excess water from the container.

Green Growers Association. How to Grow Better House Plants, *pp. 180-85. London, England. 2000. Found on home bookshelf.*

Reading 2: African Violets

African violets have beautiful flowers and heart-shaped fuzzy leaves. The flowers may be white, pink, or violet. Generally, the leaves grow on slender stalks in clusters of three. The maximum size of the plant is 3 in. to 5 in. (8 cm to 13 cm) tall.

This native African plant thrives in moderate temperatures and soil rich in decayed matter or humus. African violets are very sensitive to water. The plants need to be watered from the bottom with lukewarm water. Excess water in the saucer should be removed immediately to prevent rotting. Although the leaves can withstand a slight misting, a complete wetting of the leaves and stems will also cause rotting.

Like most plants with fleshy stems and leaves, African violets can be grown from leaf cuttings. To grow them, select a fully developed leaf and cut through the stem about 2 to 3 cm below the leaf base. Insert the stem and base of the leaf into the soil. Within several weeks, new plants should appear at the base of the leaf. When the leaves are about one-half the size of the original leaf, you can separate and transplant them.

When starting new plants, you need a high humidity environment. You can easily make one with a small paper or soft plastic container, plastic bag, sand, and vermiculite. Begin by punching several holes in the bottom of a container such as a cut-off paper milk carton, salad, sandwich, or deli dish, or paper bowl. Fill the container with a rooting mixture made from 1 cup sand and 1 cup vermiculite. Add enough water to thoroughly dampen the soil. Turn the mixture several times with a spoon. Be sure no water is standing in the container. Put 3 to 4 of your African violet cuttings in the container. Then, place the container into a plastic bag. Inflate the bag by blowing into it; seal with a rubber band or twist-it.

Excerpt from John R. Green. Growing African Violets. *The Gardener's Magazine. Pages 203-07. New York. Gardening Associated Press. 2002. Volume XI, Number 6. Reference Section. Grayson Middle School Library.*

Reading 3: Fertilizer

To grow well, plants need many different nutrients. Micronutrients or trace elements needed by plants include iron, zinc, manganese, copper, chlorine, boron, and molybdenum. Secondary nutrients include sulfur, calcium, and magnesium. Primary nutrients include nitrogen, phosphorus, and potassium or potash. The primary nutrients are the three major nutrients of fertilizer. When you buy fertilizer, you will see three numbers on the label, such as 6-8-10. These numbers are, in order, the percentages of nitrogen (6%), phosphorus (8%), and potassium (10%). Nitrogen is essential for foliage and stem growth and contributes to the dark green color of the leaves. Phosphorus encourages bloom and root growth. Potassium is needed for disease resistance and stem strength. There are many different fertilizers. Those designed for flowering plants generally contain less nitrogen because this nutrient encourages foliage growth at the expense of flowering.

Fertilizers are available in many different forms—water soluble pellets, powders, liquids, dry tablets, and sticks. You can also buy time-release pellets. Given the variety of choices, it is wise to ask the advice of a reliable plant seller who can recommend the best fertilizer for your plant's needs. The seller can also advise you on the best amount and method of application.

Excerpt from Life-Grow Chemical Company. House Plants and Fertilizer. *Pamphlet from Jones City Gardening Center. 2004. Found in Mrs. Likers' science classroom.*

Reading 4: Duckweed

Duckweeds are found on the surface of quiet freshwater ponds, streams, and lakes. All duckweeds belong to the family *Lemnaceae*. All duckweeds have leaf-like stems called fronds that float on the water's surface. Below the fronds are thread-like roots that trail in the water. All plants produce flowers.

Duckweeds can be easily collected from ponds and grown in aquaria that are well-lighted and held at average room temperature. Good lighting conditions can be produced with a fluorescent light or in front of a north-facing window. The plants can provide cover for fish and protect them from too much light. Duckweeds also absorb some nitrogenous waste produced by the animals. Contrary to popular belief, duckweeds add little oxygen to an aquarium. Once established, duckweeds may reproduce rapidly and become a problem. The problem can be solved by skimming unwanted plants from the surface with a dip net.

Although duckweeds have flowers, they rarely reproduce by flowering in aquaria. Instead, they reproduce by vegetative means. The new fronds arise from growth pockets near the sides of the parent frond. The offspring show handedness, that is, all offspring from a parent will emerge from the same pocket, either right or left. Generally, the daughter fronds will remain attached to the parent frond for a short period before separating. This growth pattern causes duckweed to cluster and form large mats on the surface of the water.

Duckweeds can also be grown in small environments such as a Petri dish, culture disk, or other small container. To make a growth medium, place a pinch of sifted loamy garden soil, 3 pellets of 8-8-8 fertilizer (one of each color), and 15 mL of distilled or aged tap water in the container. Mark your water level so you can replace the water as it evaporates. Always be sure to add aged or distilled water.

Excerpt from Duckweeds of the World. *Written by Hammer G. Carol, Louis Jones, Michael Greene, and others. Loaned by Mr. Smith, science teacher. 2003. World Botanical Society. London, England. pages 105-11.*

For more information on Exploring a Project Idea, use these search words or phrases:

"taking notes in middle school"
"referencing information"
"citing and referencing"
"citation of sources"

to find websites such as this:
standrews.austin.tx.us/library/NoteTaking.htm

Notes

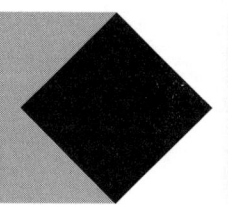

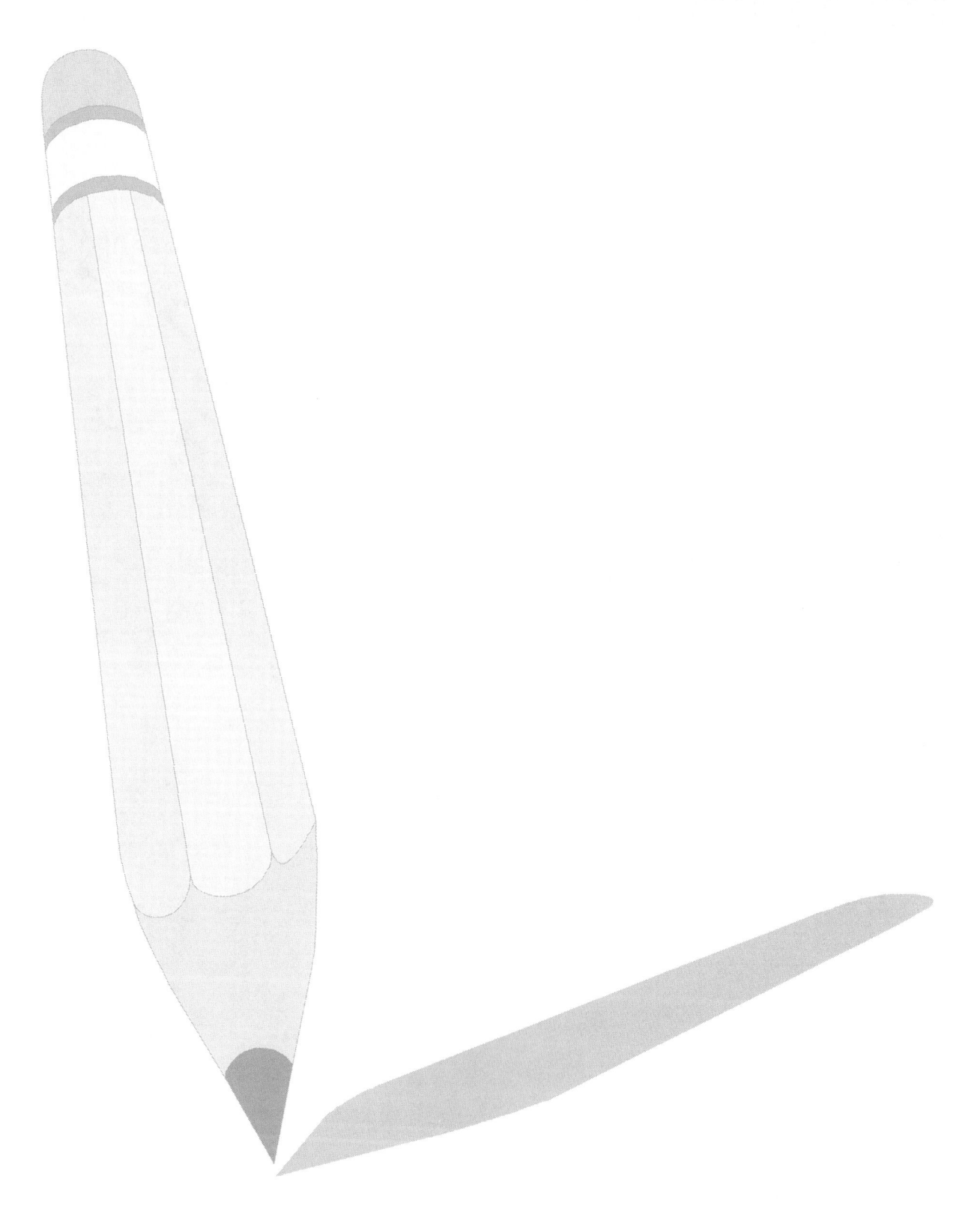

7
Writing Procedures

Materials, Steps, and Paragraphs

Does your favorite jawbreaker change colors and flavors as it dissolves? Some jawbreakers have several layers that dissolve one after another. What affects how long these candies last? What could you do to increase or decrease the dissolving time? Try **Investigation 7.1, Making the Flavors Last**, and see what you find.

Investigation 7.1

Making the Flavors Last

WHAT YOU NEED

- 2 containers
- 2 Gobstoppers™
- Cold water
- Hot water
- A watch or clock with a second hand
- Safety goggles

WHAT YOU DO

1. Get 2 identical containers.
2. Put hot water in one container and cold water in the other container.
3. Put 1 Gobstopper™ into the hot water and another Gobstopper™ into the cold water.
4. Time how long it takes for the outer layer of each Gobstopper™ to dissolve.
5. Empty and rinse the containers.
6. Did the temperature of the water affect how long it took the outer layer to dissolve?
7. Following your teacher's instructions, compare how your group conducted the experiment with the procedure followed by another group. Did they do the investigation exactly as you did?

See Chapter 8, Experimenting Safely.

Your group probably did not follow the directions in exactly the same way as another group. Because of these variations and several other reasons, this procedure needs to be improved to answer the question: "Does changing the temperature of water affect how long it takes to dissolve the outer layer of a Gobstopper™?"

Perhaps the best way to improve this procedure is to start at the beginning. Before writing a procedure, you would probably start with an idea for an experiment. Let's do that. In the process, see if you can find the flaws in the procedure above. Begin with the topic of Gobstoppers™ and use the Four Question Strategy.

Step 1: Brainstorming Topics with the Four Question Strategy

1. **What materials are readily available for conducting experiments on (*Gobstoppers™*)?**

    Possible Answers:
    Gobstoppers™
    water
    containers

2. **How do (*Gobstoppers™*) act?**

    Possible Answers:
    They dissolve.
    They change color.
    They change flavors.
    They becomes dull on surface.
    They color the water.

3. **How can you change the set of (*Gobstopper™*) materials to affect the action?**

    Possible Answers:

Gobstoppers™	Water	Containers
brand	amount	depth
size	temperature	width
age	pH	shape
number	movement	material
temperature	kind	flatness of bottom
color	substitutes	

4. **How can you measure or describe the response of (Gobstoppers™) to the change?**

    Possible Answers:
    time it takes to dissolve one or more layers
    time it takes to dissolve completely
    number of colors observed
    distance colors move in water
    pattern of dissolved colors in water

Step 2: Selecting Experimental Components

Based on your responses to the Four Question Strategy, you must now make six decisions or selections.

You must select the:	Sample selection
1. Independent Variable (from Question 3)	temperature of the water
2. Levels of the Independent Variable to be tested	water at 5, 15, 25, 35, and 45°C
3. Dependent Variable (from Question 4)	time (sec) that it takes to completely dissolve the outer layer
4. Constants (All the responses to Gobstoppers™ to Question 3 except the independent variable must be kept constant.)	same brand and color 1.2 cm in diameter 1 Gobstopper™ per container Gobstoppers™ from the same box 200 mL water 500 mL (16oz) plastic supermarket deli containers same color
5. Control	container of room temperature water
6. Number of Repeated Trials	4

Writing Procedures **77**

Do these decisions look familiar? They should! They are the major parts of an experiment. Remember the experimental design diagrams in Chapter 3? Based on these decisions you can easily write the other parts—a title and hypothesis.

Stop and summarize your selections with an experimental design diagram (compare your selections with ours in Figure 7.1). Then use the checklist in Chapter 3 to check the design of your experiment. Be sure to make any changes that will improve your experiment before proceeding to the next step.

FIGURE 7.1 EXPERIMENTAL DESIGN DIAGRAM FOR A RATE OF DISSOLVING EXPERIMENT

Title: The Effect of Water Temperature on the Time It Takes to Dissolve the Outer Layer of a Gobstopper™

Hypothesis: If the temperature of the water is increased, then the outer layer of a Gobstopper™ will dissolve faster.

IV: The temperature of the water (degrees Celsius)				
5°C	15°C	25°C (Control)	35°C	45°C
4 Trials	4 Trials	4 Trials	4 Trials	4 Trials

DV: Time (sec) for the outer layer of Gobstopper™ to dissolve

C: 1 Gobstopper™ per container
1.2 cm diameter
new box
room temperature
green

Water   250 mL
pH 7.0
no movement

Container
500 mL
7.5 cm deep
11 cm diameter
round
plastic
bottom raised center

78   Chapter 7

The choices you made for your responses to the Four Question Strategy describe just one of the many experiments that you could do. Other choices would have resulted in other experiments such as the following two.

Selections — **Alternate Experiment 2**

1. Independent Variable — Color of Gobstopper™
2. Dependent Variable — Distance dye moves in water
3. Constants — Water, container, and Gobstopper™ size, brand, age, temperature, and number used
4. Control — Red
5. Number of repeated trials — 6
6. Levels of IV tested — Red, yellow, purple, green, and orange

Alternate Experiment 3

1. Independent Variable — The concentration of sugar in water
2. Dependent Variable — Time to dissolve to the core
3. Constants — Containers, Gobstoppers™, and water, amount of, temperature, pH, movement, kind
4. Control — 0 mL sugar
5. Number of repeated trials — 5
6. Levels of IV tested — 0 mL, 75 mL, 150 mL, 225 mL, and 300 mL of sugar dissolved in 250 mL water

When making your decisions, you can only choose one independent variable. In Investigation 7.1 the independent variable stated was the temperature of the water. A major problem with the original procedure was that there were not enough constants specified. The amount, kind, pH, and movement of water were not

Writing Procedures

specified nor were the brand, size, age, temperature, or color of the Gobstoppers™. Hence, each group may have been doing a different experiment. Also, there were no repeated trials or control!

Step 3: Imagining a Sequence

Close your eyes and imagine you're performing the experiment. Then, open your eyes and jot down phrases or short sentences describing what should be done. For example:

1. Get 5 identical bowls.
2. Put 5°C water in a bowl.
3. Place a Gobstopper™ in the bowl.
4. Time how long it takes for the outer layer to completely disappear.
5. Repeat Steps 1-4 using 15°C, 25°C, 35°C and 45°C water.
6. Repeat Steps 1-5 three more times for a total of 4 repeated trials.

Step 4: Refining the Procedure

In Step 2 you made decisions about the experiment including all the things you needed to keep constant. Look at the experimental design diagram and add specified constants to the steps you visualized.

1. Get a **500 mL round (approximately 11cm diameter × 8 cm high), plastic 'deli' container** with a **ridged bottom**.
2. Put **250 mL** of 5°C water in the bowl.
3. **Carefully lower a green room temperature** Gobstopper™ (diameter = 1.2 cm) into the bowl of water.
4. Time **in seconds** how long it takes for the outer color/layer to disappear.
5. Repeat Steps 1-4 using 15°C, 25°C, 35°C and 45°C water.
6. Repeat Steps 1-5 three more times for a total of 4 repeated trials.

Now read and do Steps 1-6 of your experiment after you receive your teacher's approval and check for safety (see Chapter 8). For each step, ask yourself: Have I described this step clearly so that there is only one way to do it? Ask another group to read and then conduct 1 trial of your experiment. If they ask any questions or do anything other than what you thought that your procedure said to do, stop and fix your procedure.

Several problems related to investigating the process of dissolving Gobstoppers™ that you might encounter are described below.

Changing Temperatures

Small amounts, like 250 mL, of cold or warm water may change temperature during the time it takes to set-up and conduct an experiment. One way to solve this problem is to set each bowl of water in a larger container of water that is at the desired temperature of the water. This will not stop the change in temperature of the water in the bowl, but it will slow it down considerably.

Start Time

If you run all five repeated trials at once, the start time will be a problem. You can solve this problem with the help of an assistant or two. Have other students in your group hold the Gobstoppers™ just over the water surfaces in the bowls and gently release them when you say "Release" and start timing the process

End Point

Timing when a process is complete requires that you make a judgment call. In this case you must determine when the last speck of the outer layer of the jawbreaker disappears, and you must be consistent by making that decision at the same time in the dissolving process for each trial. This problem is solved with practice. Before you start collecting data to use in your experiment, do a few practice trials. Put a Gobstopper™ in water and observe it. Determine what the Gobstopper™ looks like as it approaches the endpoint. Then time the process until you are consistent in identifying and measuring the endpoint. Again, having a classmate check you will improve your consistency.

Comfortable Cushions

Murphy's Law states "If something can go wrong, it will, and at the worst possible time." So plan ahead. Suppose a box of Gobstoppers™ contains 22 Gobstoppers™ of each color and that you plan to do a total of 20 trials using the same color jawbreakers. Do not buy just one box of the jawbreakers, thinking you'll practice with two of the jawbreakers and use the other 20 as trials. Buy two boxes at the same store at the same time. Then you have a cushion of extra jawbreakers of the same age, production batch, and color, In fact, having a few "extras" of each of your supplies is a good idea.

The Control & Repeated Trials

You haven't been wondering about using 25°C as the control have you? Twenty-five (25°C) is about room temperature. It means you probably don't have to heat or cool the water. You just use water that has been sitting in a container for a while.

The advantage of conducting the procedure with all levels of the independent variable (Step 5) and then repeating the experiment (Step 6) is that the control is tested throughout the experiment. This allows you to better detect the effects of unwanted changes in the experiment over time. Also, the experimenter is less likely to get bored. You'll stay more alert, because the data will probably be different for each level tested.

However, sometimes it takes lots of time to change the apparatus to test the different levels of the independent variable. In that case, you might switch the order of Steps 5 and 6.

By changing your procedure to address these and other questions, you can be sure to have a clearer, more precise set of directions that can be easily and exactly followed.

Lists versus Paragraphs

Some science competitions will not allow you to write your materials as a separate list or write your procedure as a series of numbered steps. You must write your procedure in paragraph form that also includes your materials. That's O.K. No problem. Well, it's not much of a problem. You still can do everything you've learned to do to write a procedure. Because lists are easier to write than are paragraphs, first write your materials and procedures as lists, and then rewrite the lists in paragraph form. Examples of both formats are shown in Figures 7.2 and 7.3.

FIGURE 7.2 LIST FORMAT

Materials
- 5 liters of water
- 2 boxes of Gobstoppers™
- 5 identical 500 mL plastic bowls
- 1 clock or watch with a second hand
- 1 Celsius thermometer
- 1 heating device
- 1 cooling device

Steps
1. Get a 500 mL round plastic bowl.
2. Put 250 mL of 5°C water in the bowl.
3. Carefully lower a green room-temperature Gobstopper™ into the bowl of water.
4. Time in seconds how long it takes for the outer color/layer to disappear.
5. Repeat Steps 1-4 using 15°C, 25°C, 35°C and 45°C water.
6. Repeat Steps 1-5 three more times for a total of 4 repeated trials.

FIGURE 7.3 PARAGRAPH FORMAT

Place 250 mL of water into a 500 mL round plastic bowl. Carefully lower a green Gobstopper™ into the bowl of water. Using a clock with a second hand, measure the time (sec) it takes for the outer layer (color) of the Gobstopper™ to dissolve. Record the time in seconds. Rinse the bowl thoroughly.

Repeat the procedure using water at 15°C, 25°C, 35°C, and 45°C. Then repeat the experiment 3 more times for a total of 4 repeated trials.

Safety, Animals & People

When you've written the best procedure that you can write, read Chapter 8. Have your teacher check your procedure for safety. DO NOT START your experiment in any way, even to gather the necessary materials, until your procedure has been checked for safety.

If your experiment involves animals, then you need to get special forms from your teacher. An adult who is experienced in caring for animals, such as a veterinarian, teacher, or animal researcher, must approve of your procedure *before* you obtain any animals. The adult who approves your procedure must also observe you as you conduct your experiment to be certain that you do not hurt or stress the animals. The adult must also supervise your disposing of the animals when your experiment is over.

Similarly, if you plan to involve humans in your experiment, you must have the experiment approved by your teacher and the Human Subjects Committee, if your school has one, before you begin to recruit subjects or volunteers. The regulations that you must meet to do research on animals or humans are similar to the regulations that scientists must meet if they do such experiments. They must submit their procedures for approval to committees who review them to prevent unnecessary stress and suffering.

Practicing Your Skills

That's it. Well, that's almost it! Before you try to write a procedure for an experiment, you should practice writing procedures for something you already know how to do, such as:

- mowing a lawn
- dusting or polishing a table top
- making a glass of lemonade
- sharpening a pencil.

Write a procedure for one of the above. Follow the steps outlined in this chapter to improve your procedure writing skills. When you've done this, you'll be ready to write a procedure for your science experiment. Be sure to check the completeness and safety of the procedures you write.

Searching the Web

For more information on Writing Procedures, use these search words or phrases:

"scientific writing"
"writing procedures"

to find websites such as this:
an.psu.edu/jxm57/irp/mat&m.htm

Notes

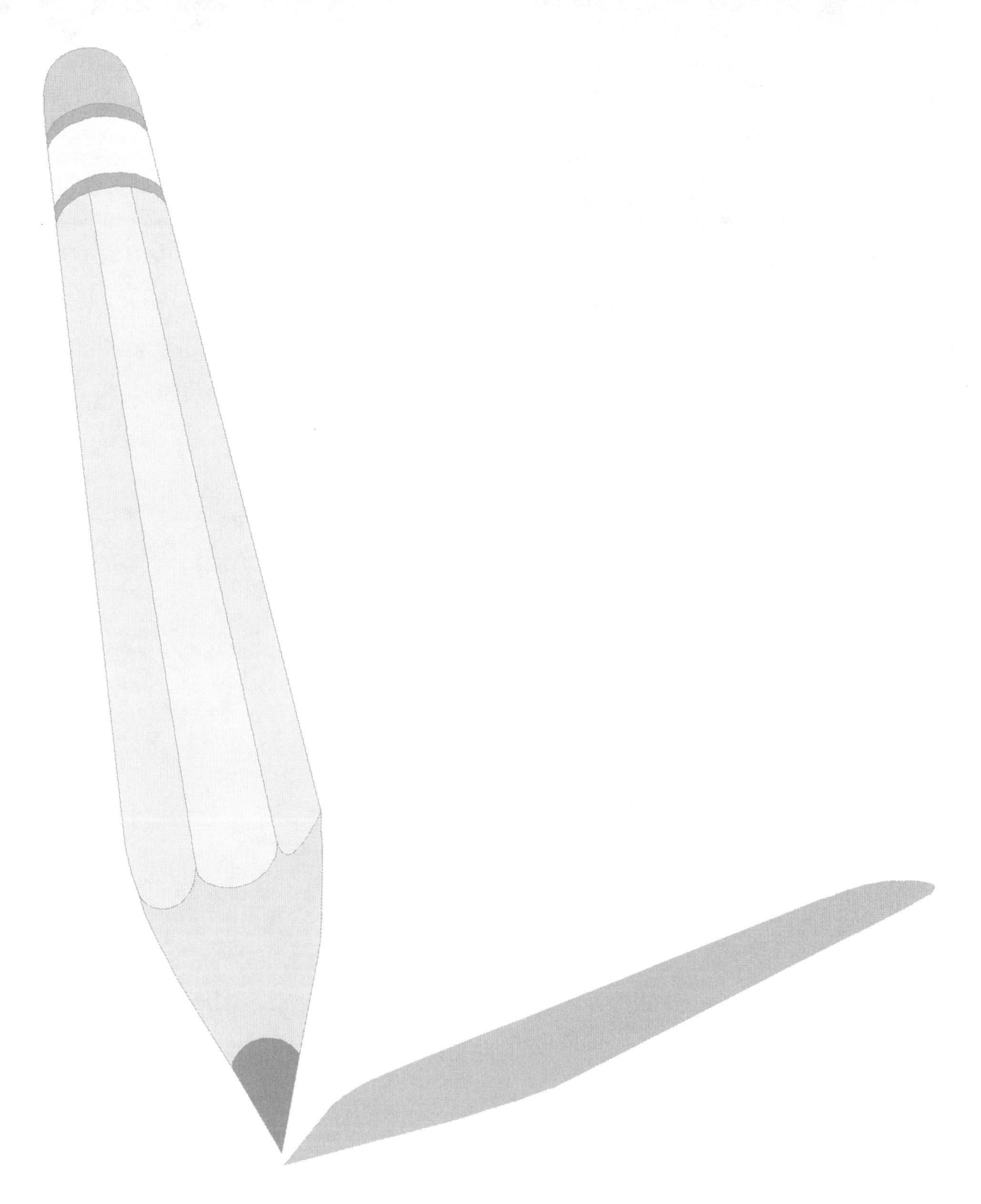

Experimenting Safely

Organisms, Chemicals, Shocks, and Radiation

Wearing your seat belt in a car and using protective pads and a helmet when skateboarding make good sense. Similar safety precautions are also important when conducting a science project. **Before you begin your experiment, be sure your teacher has reviewed your procedures for safety.** If you are conducting your experiment at home, you should also discuss your safety precautions with your parents as well.

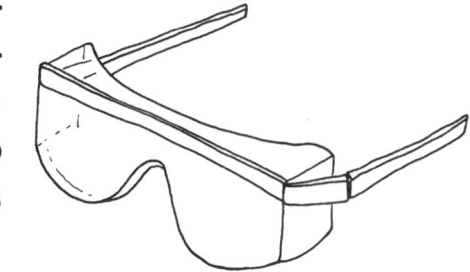

Safety concerns for different kinds of projects are described in separate sections of this chapter. These sections are:

1. chemicals,
2. mold and microorganisms,
3. electricity,
4. radiation, and
5. animals and humans.

Read the sections that are related to your project. The safety guidelines given here are only a sample. Be sure you understand and follow all the safety procedures needed for your own project.

A. Chemicals

Cleaners, fertilizers, and other chemicals serve many useful purposes, but all of them can be dangerous if improperly used. Never mix chemicals, not even household cleaners, without help from an adult. In addition, you should:

◆ **Always wear protective glasses.**
Gloves and an apron are also good ideas.

◆ **Wash your hands after handling any chemical.**

◆ **Know the potential dangers of the chemical you are using.**
Some chemicals can irritate your skin, while others are poisonous.
Do not breathe in vapors from chemicals. Be sure the area in which you are working is well ventilated.

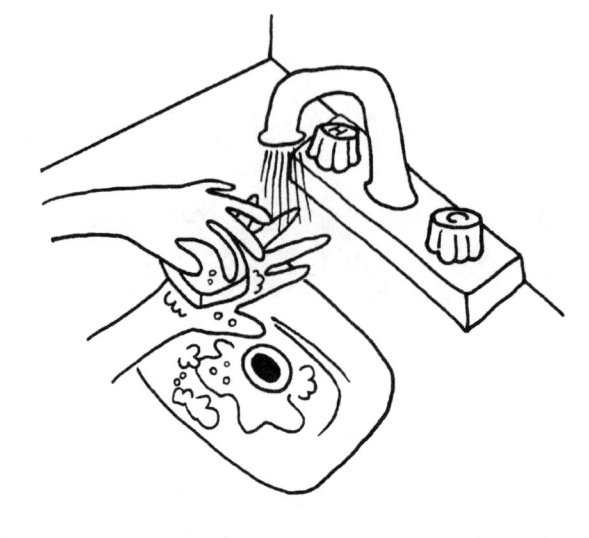

◆ **Know how and where to store chemicals safely.**
A special kind of container might be needed, or maybe the chemical should be stored in a glass instead of a plastic container.

◆ **Know what to do in case of an accident.**

◆ **Know the procedures for safely disposing your chemicals.**

Your science teacher can help you find answers to safety questions in laboratory manuals or chemical catalogs, such as the *Flinn Chemical Catalog and Reference Manual*. Most schools also have information sheets on the chemicals used in science classes. These are called Material Safety Data Sheets, or MSDS.

B. Mold and Microorganisms

You have probably seen mold growing on bread and other foods because mold spores are all around us. Microorganisms are also everywhere. Most common molds and microorganisms are harmless, but some are harmful. Before beginning any project with molds, ask your parents if you or anyone in your family is allergic to molds. Follow these safety precautions:

◆ **Keep the mold and microorganisms containers sealed.**

◆ **Do not touch the molds or microorganisms.**

◆ **Wash your hands frequently.**

◆ **Never smell molds and microorganisms by inhaling close to the containers.**

- **Do not re-use containers.**
- **Dispose of your organisms and closed containers properly.**
- **Avoid growing molds on soil, some can make you sick.**

When growing molds and microorganisms, you will often grow "uninvited" molds, bacteria, fungi, and yeasts. Most of these are also harmless, but some are not. Play it safe. Properly dispose of these uninvited guests.

Similar care should be taken when studying other microorganisms such as bacteria, protozoa, and algae. Learn as much as you can about these organisms before beginning any experiment. Bacteria, for example, are often grown in special containers called petri dishes. Harmful bacteria as well as safe bacteria may grow in these containers. Follow the same safety procedures as those given for working with molds.

C. Electricity

Experiments that use electricity should always be checked by an adult who knows how to safely work with electricity. Take the proper precautions to prevent an accident. When designing and conducting your experiment, you should:

- **Use as little voltage as possible.**
- **Avoid using current from household outlets; use batteries instead.**
- **Watch for leaky batteries. The chemicals inside can be harmful.**
- **Make sure electrical appliances and tools are insulated and grounded.**
- **Never work alone.**

D. Radiation

Experiments using microwave ovens, lasers, radon, and some types of smoke detectors all involve radiation—energy or streams of particles given off by atoms. Radiation can be very dangerous. Even in small amounts it can be harmful to living tissue.

Before beginning any experiment involving radiation, get help from someone who knows about the kind of radiation you would like to use in your experiment. Keep these safety precautions in mind:

◆ **Never work alone.**

◆ **Dispose of materials that give off radiation as required.**

◆ **Know the law. Certain state and federal laws may apply.**

E. Animals and Humans

If you plan to experiment with animals with backbones (vertebrates), you must follow very special rules. Vertebrates include fish, amphibians, reptiles, and birds as well as mammals. If you want to use vertebrates or their eggs, discuss your ideas with your teacher first. A qualified adult supervisor who is trained to take care of vertebrates, like a zoologist or a veterinarian, must agree to begin supervising your project before you even obtain the first organism. A form like the one in Figure 8.1 must be signed by the person agreeing to supervise a project involving vertebrates.

Most schools and competitions very strongly recommend that students use animals without backbones (*in*vertebrates) in animal experiments. Insects and worms are examples of invertebrates. If you do a project with animals, you must provide proper care for all the animals. Proper care includes:

◆ **A comfortable living place;**

◆ **Procedures that do not injure the organism;**

◆ **Enough food, water, warmth and rest;**

◆ **Gentle handling;**

◆ **Humane disposal or a proper home for organisms when your experiment is finished.**

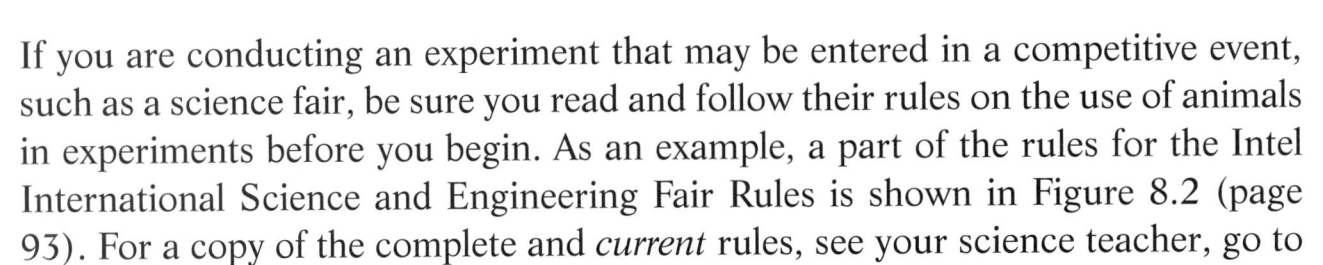

If you are conducting an experiment that may be entered in a competitive event, such as a science fair, be sure you read and follow their rules on the use of animals in experiments before you begin. As an example, a part of the rules for the Intel International Science and Engineering Fair Rules is shown in Figure 8.2 (page 93). For a copy of the complete and *current* rules, see your science teacher, go to

FIGURE 8.1 REQUIRED SUPERVISION FORM WHEN CONDUCTING VERTEBRATE ANIMAL RESEARCH

Non-Human Vertebrate Animal Form (5)
Required for all research involving nonhuman vertebrate animals.
(SRC approval required before experimentation.)

ATTENTION: *This form is not necessary if student uses only tissue from non-human vertebrates in the project.*

Student's Name _____

Title of Project _____

To be completed by Student Researcher:

1. Genus, species, common name of animal(s) used. (**Use separate animal form for each species used.**)

   _____

2. Where will animals be obtained? (See p. 17); Pet store animals, except fish and those used for behavioral studies, are inappropriate for research.

   _____

3. How many animals will be used? _____ Average weight _____

4. Cage size _____ Number of animals per cage _____

5. Type of food _____

6. How often fed and given water? _____

7. Type of bedding used (Do not use cedar chips, newspaper, or paper towels.) _____

8. Where will animals be housed? _____

9. Name the veterinarian who will provide veterinary medical and nursing care in case of illness or emergency (**required**).
   D.V.M. _____ Name of Facility _____ Phone _____

10. Will animals be euthanized? ☐ Yes   ☐ No

    If yes, why and by what method? _____ By whom? _____

    If no, what will happen to the animals after experimentation? _____

To be completed by Animal Care Supervisor:

Name _____

Position _____

Institution _____

Address _____

Office Phone _____

I certify that I have discussed this research with the student prior to its start and will supervise and will accept primary responsibility for the quality of care and handling of the live vertebrate animals used by the above named student. I further certify that I am knowledgeable in the proper care and handling of laboratory animals, and meet prevailing animal care supervisory requirements. When an animal must be euthanized, I certify that I will perform the procedure, using recommended agents.

Animal Care Supervisor's Printed Name _____ Signature _____ Date of Approval
(Must be prior to experimentation.)

Title _____ Phone _____

Institution and Address _____

International Rules 2003/2004 full text of the rules and electronic copies of forms are available at www.sciserv.org/isef   Page 38

Science Service, Incorporated. Rules of the 2004 International Science and Engineering Fair. *Washington, DC: Science Service, Inc., 2003.*

Experimenting Safely   **91**

www.sciserv.org/isef/document/index.asp and click on Rules and Regulations, or use a search engine and the entry "Intel International Science and Engineering Fair."

Special rules must also be followed in experiments using humans. Nothing may be done to humans that is likely to cause them harm. Participation should be voluntary. Some experiments, like those that just involve observing people, may not need special signed forms and procedures. Talk with your teacher about experiments involving humans before you begin. Scientists who wish to do experiments on humans or animals must have their research plans approved by a committee of fellow scientists. These rules are to help insure that humans and animal subjects are treated properly.

Summary

There are risks with everything we do. Taking proper precautions and using safe procedures can reduce these risks. Cooking, for example, can be dangerous. But you can cook safely by being careful and following safety procedures that reduce the danger. That's why people use potholders and keep pot handles pointed in toward the stove. When you conduct your science experiment, practice good safety procedures, too. Safety is no accident, plan for it.

For more information on Experimenting Safely, use these search words or phrases:

"lab safety rules"
"science fair safety rules"
"experimenting safely"

to find websites such as this:
hants.gov.uk/school/hawleyprimary/Policies/science.htm

FIGURE 8.2 SAMPLE RULES FROM THE INTEL INTERNATIONAL SCIENCE AND ENGINEERING FAIR

Display and Safety Regulations

Not Allowed at Project or in Booth

1. Living organisms, including plants
2. Taxidermy specimens or parts
3. Preserved vertebrate or invertebrate animals
4. Human or animal food
5. Human/animal parts or body fluids (for example, blood, urine) (Exceptions: teeth, hair, nails, dried animal bones, histological dry mount sections, and completely sealed wet mount tissue slides)
6. Plant materials (living, dead, or preserved) which are in their raw, unprocessed, or non-manufactured state (Exception: manufactured construction materials used in building the project or display)
7. Laboratory/household chemicals including water (Exceptions: water integral to an enclosed apparatus or water supplied by the Display and Safety Committee)
8. Poisons, drugs, controlled substances, hazardous substances or devices (for example, firearms, weapons, ammunition, reloading devices)
9. Dry ice or other sublimating solids
10. Sharp items (for example, syringes, needles, pipettes, knives)
11. Flames or highly flammable materials
12. Batteries with open-top cells
13. Awards, medals, business cards, flags, acknowledgments, etc. (Exception: The current year Intel ISEF medal may be worn at all times.)
14. Photographs or other visual presentations depicting vertebrate animals in surgical techniques, dissections, necropsies, or other lab procedures
15. Active Internet or e-mail connections as part of displaying or operating the project at the Intel

Maximum Size of Project at the Intel ISEF

30 inches (76 centimeters) deep
48 inches (122 centimeters) wide
108 inches (274 centimeters) high from floor to top of project

At the Intel ISEF, fair-provided tables will not exceed a height of 36 inches (91 centimeters).

Project must be positioned at the back of the booth and parallel to the rear of the booth.

Allowed at Project or in Booth BUT with the Restrictions Indicated

1. Soil or waste samples **if permanently encased in a slab of acrylic**
2. Postal addresses, World Wide Web and e-mail addresses, telephone numbers, and fax numbers **of Finalist only**
3. Photographs and/or visual depictions **if:**
   a. Credit lines of their origins: "Photograph taken by..." or "Image taken from..." are attached. (If all photographs being displayed were taken by the Finalist, one credit line prominently displayed indicating that the Finalist took all photographs is sufficient.)
   b. They are from the Internet, magazines, newspapers, journals, etc., and credit lines are attached.
   c. They are photographs of the Finalist or the Finalist's family
   d. They are photographs of human subjects for which signed consent forms are at the project or in the booth
   e. They are not deemed offensive by the Scientific Review Committee, the Display and Safety Committee, or Science Service.
4. Any apparatus with unshielded belts, pulleys, chains, or moving parts with tension or pinch points **if for display only and not operated**
5. Class II lasers **if:**
   a. Operated only by the Finalist
   b. Operated only during Display and Safety inspection and during judging
   c. Labeled with a sign reading "Laser Radiation: Do Not Stare Into Beam"
   d. Enclosed in protective housing that prevents physical and visual access to beam
   e. Disconnected when not operating
6. Class III and IV lasers **if for display only and not operated**
7. Large vacuum tubes or dangerous ray-generating devices **if properly shielded**
8. Empty tanks that previously contained combustible liquids or gases **if certified as having been purged with carbon dioxide**
9. Pressurized tanks that contained non-combustibles **if properly secured**
10. Any apparatus producing temperatures that will cause physical burns **if adequately insulated**

Science Service, Incorporated. Rules of the 2004 International Science and Engineering Fair. *Washington, DC: Science Service, Inc., 2003.*

Experimenting Safely 93

Notes

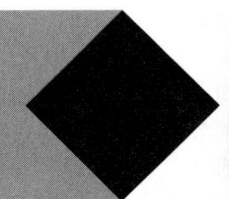

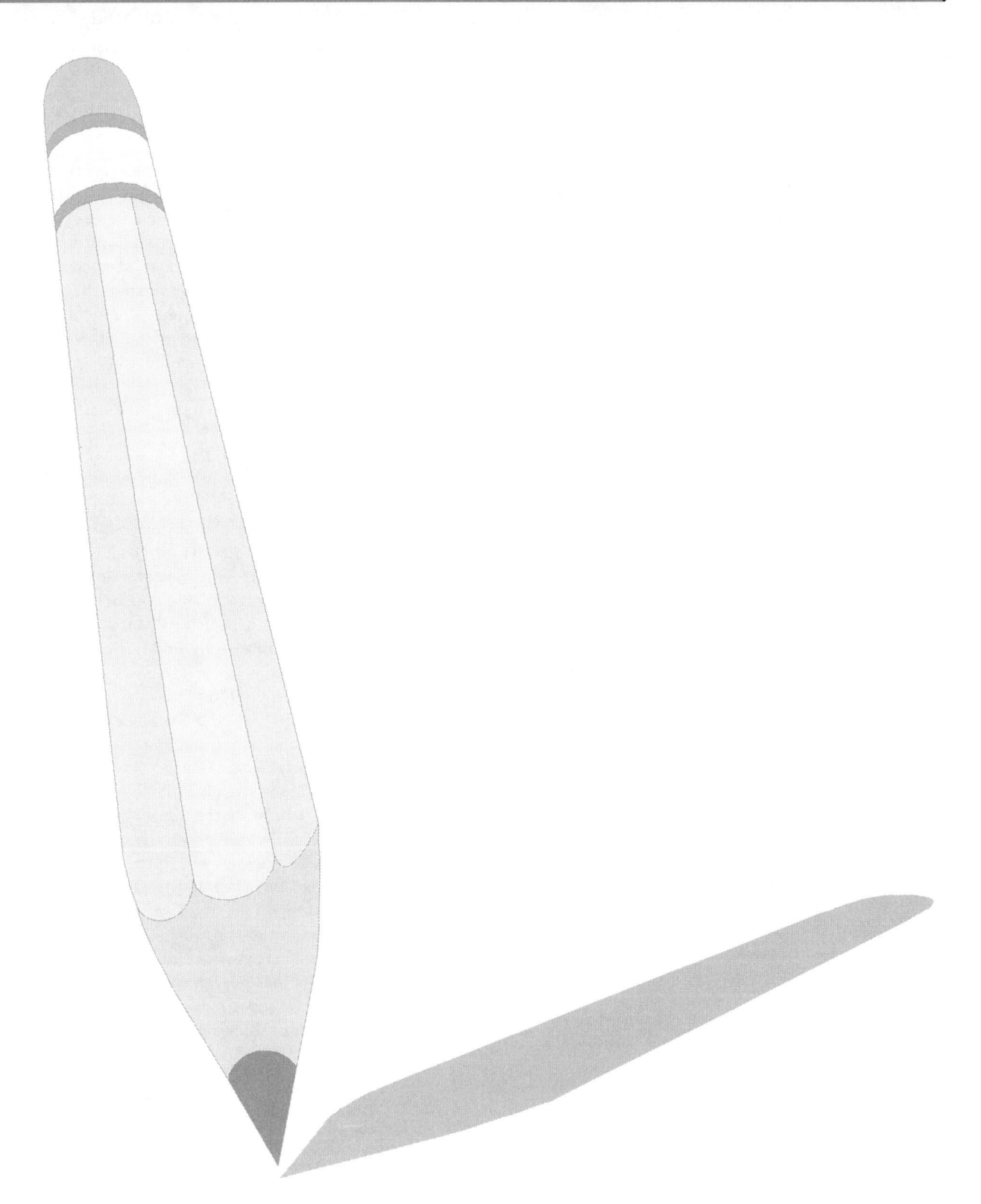

Recording Data

Tables and Data Summaries

Making a data table is a lot like setting the table for dinner. Both require a plan for organizing—on which side of the plate does the fork go? Is the independent variable always on the left side of a data table?

Organizing data from experiments into tables helps scientists better understand their results. Just as there are no absolute rules about setting a dinner table, "Miss Manners" not withstanding, there are no rigid rules for constructing a data table. However, in both cases there are commonly agreed upon patterns of organization that make it easier to know whose fork, yours or mine, and which variable, independent or dependent, is on the left.

Stating Hypotheses

Before conducting experiments and collecting data, scientists usually state a hypothesis that predicts how two or more variables are related. After data are collected, the data are used to either support or not support the original hypothesis. You learned how to write a hypothesis using an **If..., then...** sentence in Chapter 1. When you conduct your own experiment, making a data table will make it easier to see if your data support or do not support your hypothesis.

Parts of a Data Table

Before collecting data from your experiment, you should first plan ways to organize and display your data in the form of a chart called a **data table**. Although there are no absolute rules for constructing a data table, there are general guidelines to help you (see Table 9.1).

For example, the independent variable is almost always recorded in the left column and the dependent variable in the middle column. When repeated trials are conducted, the middle column is divided into smaller columns. The number of smaller columns should be equal to the number of repeated trials. Additional information, such as a central or typical value (**mean, median,** or **mode**) or the **range**, is recorded in the column to the right of the dependent variable column. Information you figure out from data, like a mean, is called a **derived quantity.**

When recording data in a table, the values of the independent variable are ordered. These are usually ordered from smallest to largest. Organizing the data in this way creates a pattern of change in the independent variable. If there is any pattern of change in the dependent variable, it will be easier to see if the levels of the independent variable are put in order.

When labeling the columns in a data table, include the units of measurement in parenthesis, for example, Time to Evaporate (seconds). You can also use abbreviations like (sec). The title of the data table should clearly communicate the information contained in the table. The variables that were investigated are usually included in the title, such as "The Effect of Number of Alcohol Drops on Evaporation Time." Notice that all major words in the title are capitalized. The Effect of the (*Independent Variable*) on the (*Dependent Variable*) is an easy way to write a title.

Suppose you wanted to conduct an experiment to see if the number of drops of rubbing alcohol (1, 2, 3, and 4 drops) affected the time it took for the alcohol to evaporate. The number of drops of alcohol would be your independent variable, while time to evaporate would be your dependent variable. If you tested each level of the independent variable three times, you would divide the column for the dependent variable into three smaller columns when making your data table.

Using the guidelines for making a data table, you would construct a table like Table 9.1.

If you would like to try the experiment, follow the steps in Investigation 9.1.

TABLE 9.1 The Effect of Amount of Alcohol Drops on Evaporation Time

Amount of Alcohol (drops)	Time to Evaporate (sec) Trials			Mean Time to Evaporate (sec)
	1	2	3	
1				
2				
3				
4				
Column for the Independent Variable	Column for the Dependent Variable			Column for a Derived Quantity

Investigation 9.1

Cool Hands

What You Need

- Rubbing alcohol (isopropyl)
- Index card to use as a fan
- Pippette or medicine dropper
- Safety goggles

Wear eye goggles. See Chapter 8, Experimenting Safely.

What You Do

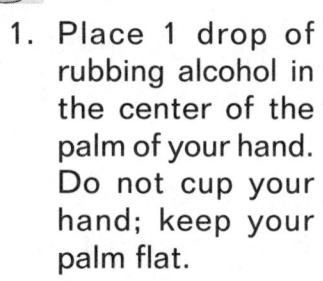

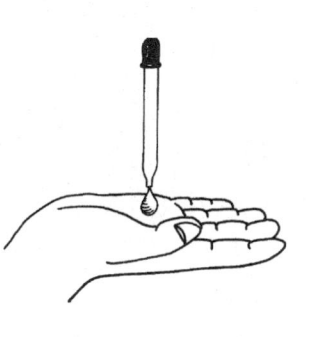

1. Place 1 drop of rubbing alcohol in the center of the palm of your hand. Do not cup your hand; keep your palm flat.
2. Time how many seconds it takes the alcohol to evaporate while fanning your hand with an index card. Record your data in the data table.
3. Repeat Steps 1 and 2 two more times for a total of 3 trials.
4. Repeat Steps 1–3 with 2, 3, and then 4 drops of alcohol.

Recording Data

TABLE 9.2 The Effect of Amount of Alcohol Drops on Evaporation Time

Amount of Alcohol (drops)	Time to Evaporate (sec) Trials 1	2	3	Mean Time to Evaporate (sec)
1	20	22	22	21
2	38	40	42	40
3	47	48	52	49
4	62	60	63	62

How do your data compare with our data in Table 9.2? Your data may be different because some of the following constants were probably not the same as ours: the size of your drops, the temperature of your hand, and how hard you fanned.

Quantitative Data

In Chapter 2 you learned about types of data. One type of data that is measured or counted is called **quantitative** data because the data are different quantities. Quantitative data can be summarized by calculating the mean and the range.

Mean

Conducting repeated trials for three drops of alcohol resulted in three different times, 47, 48, and 52 seconds. Which one of these results, called **values**, is correct or best? Why conduct repeated trials if the resulting values are not the same each time? Many measurements contain chance errors that are often unavoidable. Rulers can not always be held exactly the same; not all thermometers read exactly the same; and some timers run faster than others. No single value can be picked as perfect. The value at the center of a cluster of data is chosen as the most typical or central value. Finding the mean of the data is one method to find the most typical or central value. Chance errors that make some measurements too high and others too low are balanced when the mean is calculated.

You can summarize measurements and counts by finding the mean. The mean is also called the arithmetic average.

$$\text{Mean} = \frac{\text{Sum of measurements}}{\text{Number of trials}} \quad \text{or} \quad \text{Mean} = \frac{\text{Sum of counts}}{\text{Number of trials}}$$

The mean becomes the most typical value for the repeated trials. The mean time it took 1 drop of alcohol to evaporate in our experiment was 21 seconds:

$$\frac{20 \text{ sec} + 22 \text{ sec} + 22 \text{ sec}}{3} = \frac{64 \text{ sec}}{3} = 21.3 \text{ rounded to 21 sec.}$$

In your experiment, what was the mean amount of time for 1 drop of alcohol to evaporate?

Range

In Table 9.2 the different times it took 1 drop of alcohol to evaporate were 20, 22, and 22 seconds. Were these times close to each other or spread out? A measure of how spread out the data are is called the **range**. The range tells us how much variation there is in the data. For each level of the independent variable, you find the range by subtracting the smallest (minimum) value from the largest (maximum) value of the dependent variable. For 1 drop of alcohol, the largest evaporation time was 22 sec and the smallest time was 20 sec. The difference between the two values is 2 sec (22 sec - 20 sec = 2 sec). Compare this range to the range for 3 drops of alcohol. The values for 3 drops were 47, 48, and 52 sec. What's the range? Subtract the smallest value (47) from the largest value (52): 52 sec - 47 sec = 5 sec. For 3 drops of alcohol, the range (variation) of evaporation times was greater that the range for 1 drop of alcohol. For 1 drop of alcohol the evaporation times were more alike than for 3 drops where there was a greater difference in evaporation times. There was greater variation in evaporation times for 3 drops of alcohol than there was for 1 drop.

The range is important because sometimes experimental groups can have the same mean yet be very different. For the data group of 20, 22, and 22 seconds, the mean is 21 seconds and the range is 2. For another data group of 10, 22, and 32 seconds, the mean is also 21 seconds. This range, however, is 12 seconds! The larger range tells us that this data group is more spread out and has greater variation than the group with the smaller range.

Recording Data

Summarizing Sentences

How would you describe to another person what happens to evaporation time as the number of alcohol drops increases? A good way to summarize your data table is to explain your data by comparing the most typical values, like the means, and comparing the ranges. How are the means alike or different for the various number of alcohol drops? Which number of drops evaporated quickest? slowest? How much variation in evaporation times was there for each number of drops?

Did your data support your hypothesis? Try discussing the data with a friend or someone in your family before writing a summary of your results in a few sentences. A summary based on the **mean** time to evaporate and the **range** of that data might look like this:

> The mean time to evaporate increased as the number of drops of alcohol increased. One drop evaporated with a mean of 21 sec, while 4 drops evaporated with a mean of 62 sec. The range in evaporation time was smallest for both 1 and 4 drops of alcohol, and much larger for 2 and 3 drops of alcohol. The hypothesis stated that if the number of alcohol drops were increased, then the time to evaporate would also increase. The data supported the hypothesis.

Qualitative Data

Not all data are quantities. Some data are more like descriptions. These data are called **qualitative** data because they describe qualities, such as color or gender. Qualitative data are summarized using the mode or the median and a frequency distribution.

Mode

How did your hand feel after the drops of alcohol evaporated? Suppose you described how your hand felt after each different number of drops of alcohol evaporated. You could make a scale to describe how cool your hand felt:

VC = Very Cool
MC = Moderately Cool
SC = Slightly Cooler
NC = No Change

100   Chapter 9

When you have data that are measured, you can figure out the most typical value by determining the mean. But how can you add NC + SC + SC + MC + SC and divide by 5? When the data are qualitative or described data, like word labels, you need a different procedure for determining the most typical value. Fortunately, this is very easy. It's the one value that appears most often.

Notice in the Table 9.3 that the most frequent or typical value for 1 drop of alcohol is "SC" for slightly cooler. What would be the most typical value for 3 drops? The most typical value for described data is called the **mode**.

TABLE 9.3 The Effect of the Amount of Alcohol on Feeling Coolness

Amount of Alcohol (drops)	\multicolumn{5}{c}{Coolness Feeling Trials}	Mode
	1	2	3	4	5	
1	NC	SC	SC	MC	SC	SC
2	MC	MC	MC	MC	MC	MC
3	MC	VC	MC	MC	VC	MC
4	VC	VC	VC	VC	VC	VC

Frequency Distribution

In Table 9.3, the data for every trial for 4 drops were the same, all five trials were VC; there was no variation. For 3 drops, however, there was some variation; the data were 3 MC and 2 VC. Variation in qualitative data is described using a frequency distribution that shows the number of results for each category of a variable. A frequency distribution for the results in Table 9.3 looks like this:

		1 drop	2 drops	3 drops	4 drops
Very Cool	VC	0	0	2	5
Moderately Cool	MC	1	5	3	0
Slightly Cool	SC	3	0	0	0
No Change	NC	1	0	0	0

Notice how the distribution of coolness is different for each number of alcohol drops. A frequency distribution organizes the data to show variation in qualitative data, just as the range provides information about variation in quantitative data.

Summary Sentences

Can you describe how hands feel as more drops of alcohol are evaporated? How are the modes alike or different for the various number of alcohol drops? Which number of evaporating drops felt the coolest? Least cool? How much variation was there in how hands felt for each number of drops? Do the data support the hypothesis?

A summary using the data on the **most typical cooling effect (mode) and the variation (frequency distribution)** of the data might look like this:

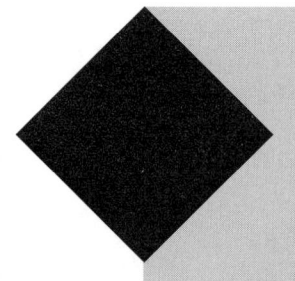

> As more drops of alcohol evaporated, the hand felt increasingly cooler. While the evaporation of one drop felt cool, the evaporation of 4 drops felt very cool. The greatest variation in how the cooling was felt was for one drop and three drops. For two and four drops, there was no variation; for two drops the rating was always moderately cool and for four drops the rating was always very cool. The data supported the hypothesis that as more drops of alcohol evaporated, the hand would feel cooler.

Median

In some experiments the data could also be ordered in a logical sequence. Three sequences often used in experiments to order the data are: smallest to largest, weakest to strongest, or worst to best. What would such an experiment look like? Well, you just did one when you conducted Investigation 9.1 Cool Hands!

Suppose you used a numerical scale like the one below to record the data you collected in Investigation 9.1, Cool Hands:

1 = No Cooling
2 = Slight Cooling
3 = Moderate Cooling
4 = Great Cooling

TABLE 9.4 The Effect of the Number of Drops of Alcohol on the Coolness Feeling

Amount of Alcohol (drops)	Coolness Feeling Trials					'Central Value'
	1	2	3	4	5	
1	1	2	2	3	2	
2	3	3	3	2	4	
3	3	4	3	3	4	
4	4	4	4	4	4	

Table 9.4 is missing a measure of 'central value.' We could determine the mode for each drop. However because the scale we used is numerical the data can be rank ordered. We can use the ordered data to determine another kind of central value, the **median**. To determine the median use the following procedure.

Step 1. Order the data.
For 1 drop the data ordered from least to greatest would look like this:

Cooling ratings when 1 drop was used:
1
2
2
2
3

Step 2. For each level of the independent variable determine the middle value or **median** of the data. Half of the data will be above the median and half will be below.

Cooling ratings when 1 drop was used:
1
2
middle value (median) → 2
2
3

Recording Data   103

Suppose you repeated the experiment, but collected 6 pieces of data each time instead of 5. When you have an even number of data, the middle value is between two numbers. Calculate the arithmetic average of these two numbers to determine the median.

Cooling ratings when 6 measurements were made:

$$
\begin{array}{l}
2\\
2\\
2\\
\end{array}
$$

median → 2.5   (2 + 3 ÷ 2 = 2.5)

$$
\begin{array}{l}
3\\
3\\
4\\
\end{array}
$$

We can now complete Table 9.4 by determining the median value for each set of data.

TABLE 9.4 The Effect of the Number of Drops of Alcohol on the Coolness Feeling

Amount of Alcohol (drops)	Coolness Feeling Trials					Median
	1	2	3	4	5	
1	1	2	2	3	2	2
2	3	3	3	3	3	3
3	3	4	3	3	4	3
4	4	4	4	4	4	4

Frequency Distribution

The variation in the data in Table 9.4 can be described through a frequency distribution that shows the different coolness ratings given for each number of drops.

	1 drop	2 drops	3 drops	4 drops
1 - No Cooling	1	0	0	0
2 - Slight Cooling	3	0	0	0
3 - Moderate Cooling	1	5	3	0
4 - Great Cooling	0	0	2	5

Summary Sentences

A summary using the data on the most typical or central value for cooling effect (the **median**) and the variation (a **frequency distribution**) would be very similar to a summary using the mode. The summary of the data might look like this:

> As more drops of alcohol evaporated, the cooling effect increased. While the evaporation of one drop resulted in a slight cooling effect, the evaporation of 4 drops resulted in a great cooling effect. The greatest variation in how much the cooling effect was felt was for one drop and three drops. For two drops and four drops, there was no variation; for two drops the rating was always 3-moderate cooling while for four drops the rating was always 4-great cooling. The data supported the hypothesis that as more drops of alcohol evaporated, the hand would feel a greater cooling effect.

Summarizing Your Own Data

Making paper worms grow is amazing! But, do all paper worms grow the same amount? When they grow, do they all have the same shape? Well, now's the time to collect some data and use your new knowledge of recording data to find out.

Save your data and straws. In Chapters 13 and 14 you'll learn new ways of graphing and analyzing the data. You will need the straws to do **Investigation 14.2, Straw Javelins** on page 106.

For more information on Recording Data, use these search words or phrases:

"quantitative and qualitative data"
"mean, median, and mode"
"frequency distributions"

to find websites such as this:
zone101.com/LearningZone/MathZones/theory/grade6/meanmode.htm

Searching the Web

Investigation 9.2

Paper Worms

What You Need

- Straw with paper wrapper
- Small container of water
- Metric ruler
- Eyedropper or pipette
- Safety goggles

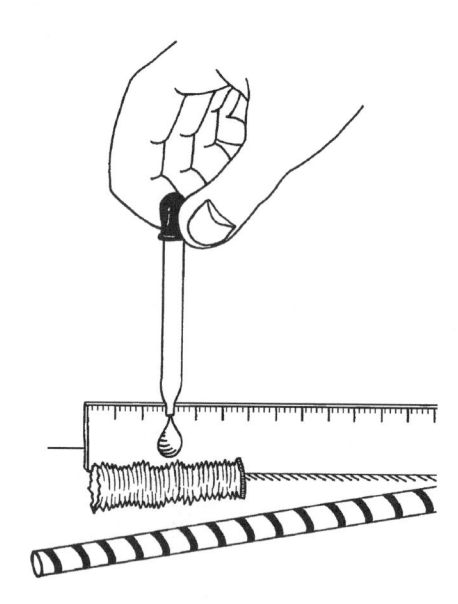

Dispose of straws as directed by your teacher. Do not put straws in your mouth. See Chapter 8, Experimenting Safely.

What You Do

1. Tear off just one end of the straw's paper wrapper. Grip the straw about 2-3 cm from the other end and tap that end on a flat surface while pushing the wrapper down into a series of small folds. Keep moving your fingers up the straw about 2-3 cm and repeat the procedure until the wrapper is pushed together at the bottom of the straw. Then remove the paper worm by pushing it off the straw.

2. If the paper worm is not straight, gently straighten it. Measure the length of the paper worm in mm.

3. Using an eyedropper (or your straw) put 4 drops of water along the length of the paper worm.

4. Use the following scale to rate the straightness of the paper worm: 1 = straight, 2 = mostly straight, 3 = curved, 4 = very curved.

5. If the wet paper worm is not straight, gently straighten it. Measure the length of the wet paper worm in mm.

6. Enter your data in a class data table. If necessary repeat Steps 1-5 so that your class has at least 25 trials.

7. Calculate the means and ranges for your quantitative data and medians and frequency distributions for your qualitative data. Write summary sentences for each.

Practice Problems

Recording Data

1. Construct a data table for an experiment on ways to keep apples from turning brown. The title of the study was, "The Effect of Preservatives Such as Lemon Juice, Lime Soda, and Fruit Freshener on the Amount of Brown Discoloration of Apples." Five trials were conducted.

2. It was hypothesized that cucumber peelings would produce more gas than other kinds of peelings. What is the mean amount of gas produced for each type of peeling? What is the range? Did the data support the hypothesis?

The Effect of Various Types of Peelings on the Production of Gas					
Type of Peeling	Amount of Gas (mL) Trials			Mean Amount of Gas (mL)	Range (mL)
	1	2	3		
Potato	40	39	41		
Apple	30	32	32		
Orange	34	32	35		
Cucumber	0	0	0		

Recording Data **107**

3. It was thought that larger crystals would grow at cooler temperatures, while smaller crystals would grow at warmer temperatures. What is the most typical crystal size grown at each temperature? How much spread is there for each temperature?

The Effect of Various Temperature Conditions on the Size of Crystals								
Temperature (°C)	Size of Crystal Trials					Mode	Frequency Distribution
	1	2	3	4	5		
10	L	L	M	L	L		
20	L	M	M	L	M		
30	M	M	S	M	S		
40	S	S	S	S	S		

VS = Very Small; S = Small; M = Medium; L = Large*

*If you measured the crystals in millimeters or fractions of an inch, you could have a mean as your most typical value.

4. Correct or improve this data table:

Time for Ice to Melt	Salt Water Concentration (g/100 mL)	Mean Time	Range (sec)
	2		
	0		
	4		
	3		
	1		

5. Write a summary of the experiment in practice problems 2 or 3.

Constructing Bar Graphs

Axes, Bars, Distributions, and Sentences

If a picture is worth a thousand words, then so is a graph. Because graphs communicate data in pictorial form, they communicate patterns or information better than a data table. Learning to construct graphs will require stretching your skills. Let's begin by investigating the behavior of a familiar stretchy object, the rubber band, in *Investigation 10.1, The Thick and Thin of It.*

Investigation 10.1

The Thick and Thin of It

WHAT YOU NEED

- 3 rubber bands (thin, medium, thick width)
- Metric ruler
- Flat surface such as a table

SAFETY: Do not shoot bands in the direction of people, animals, and objects. See Chapter 8, Experimenting Safely, Section E, Animals and Humans.

WHAT YOU DO

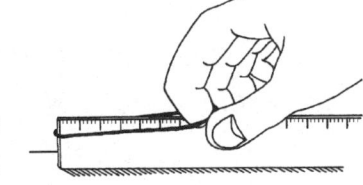

1. Hold a ruler on its edge on a flat level surface.
2. Place a thin rubber band so that one end of it just catches on the end of the ruler.
3. Pull the other end of the rubber band until the sides of the rubber band are flat against the ruler but the rubber band is not stretched. Continue to pull the rubber band until it is stretched 1.0 cm. Release it.
4. Measure in centimeters how far the rubber band moved before it hits the flat surface.
5. Repeat Steps 2-4 until you have a total of 4 trials.
6. Repeat Steps 2-5 using a medium and a thick rubber band.

TABLE 10.1 The Effect of Rubber Band Width on Distance Traveled

Width of Rubber Band	Distance Traveled (cm) Trials 1	2	3	4	Mean Distance (cm)	Range (cm)
Thin	557	564	556	579	564	23
Medium	535	548	539	535	539	13
Thick	512	508	518	522	515	14

How does your data compare with ours? Your data may be different because some of the constants probably were different in your experiment: the brand of rubber band, the width of the rubber bands, the composition of the rubber, air currents, the temperature of the rubber band and the room.

Constructing Bar Graphs

Now, let's use the data to construct a graph such as the one shown in Figure 10.1. To construct a bar graph, identify the independent variable, width of rubber band, and write it below the horizontal line of the graph (X axis). Because you used 3 different widths of rubber bands, divide the X axis into 3 equal sections or intervals and label each interval with a width (thin, medium or thick).

> **Remember**
>
> The **independent variable** is on the horizontal line, or X axis, of the graph.
>
> The **dependent variable** is on the vertical line, or Y axis, of the graph.

Write the dependent variable, distance traveled (cm), along the side of the graph (Y axis). On the graph you will use the mean distance traveled. The ranges are not displayed on the graph. The next task is to divide the Y axis into intervals that are appropriate for the data. An easy way to determine how many units each interval of the Y axis represents consists of a few steps.

◆ Find the difference between the largest and smallest numbers to be graphed (our numbers were 564 − 515 = 49).

◆ Divide the difference by number of intervals you want. If you want about 5 intervals, divide by 5. This usually results in a scale with 5-7 intervals. Too many intervals crowd a graph, while too few make it difficult to plot data points. (49 ÷ 5 = 9.8).

◆ Using 9.8 to make intervals would be too hard, so make the job easier by rounding 9.8 to a good counting number like 10. Use 10 to mark off the intervals.

◆ Because 515 is the smallest number you need to graph, you can begin with a multiple of 10 below the smallest number (510) and move upwards. Beginning the graph with 0 would waste space.

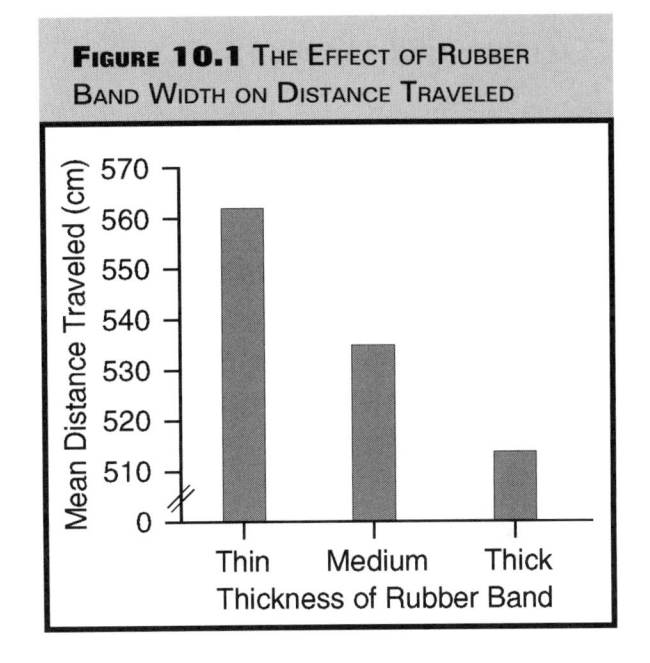

FIGURE 10.1 THE EFFECT OF RUBBER BAND WIDTH ON DISTANCE TRAVELED

Some people think that all graphs begin with the intersection of the X and Y axes labeled as (0, 0). Graphs that begin at (0, 0) work fine when the data sets start at or near zero. But in the real world and in many experiments, data sets often begin with numbers far from zero, such as in this experiment on rubber bands. Beginning a graph at 0 for data far from 0 results in a large gap between 0 and the first piece of data to be graphed. When you don't begin at 0, you can use the symbol (≠) to indicate that part of the graph is not shown. In this graph numbers below 510 are not shown.

To graph the data, draw a vertical bar from a value of the independent variable (thin) to the corresponding value of the mean dependent variable (564 cm). Make each bar the same width and leave equal spaces between the bars.

You will also need to write a summary sentence for the graph. Here's our summary.

◆ As the width of the rubber band increased, the distance traveled decreased.

Discrete and Continuous Variables

The data from this experiment were displayed as a bar graph because types of rubber bands are discrete categories. **Discrete** means that the categories are separate and do not overlap. The spaces or intervals between categories have no meaning. Examples of discrete data are the days of the week, kind of animal, gender, brands of paper towels, and types of vehicles. In your experiment you used word labels such as thin, thick, and medium to describe the rubber band widths. There was no rubber band width between medium and thick such as "medium-thick." Discrete data can only be displayed as bar graphs.

> **Remember**
>
> **Discrete** means that the categories are separate and do not overlap.
>
> **Continuous** means that the values are not separate and the intervals between them have meaning.

If you had used actual measures of rubber band widths such as 5, 10, or 15 mm, you would have continuous data. **Continuous** means that the values of the variable are not separate categories and the intervals between have meaning. For example, you could predict how far a rubber band would fly if it had a width of 8 mm or 14 mm. Other examples of continuous variables are amounts of water, weight of tomatoes, and speed of cars. Continuous variables can be displayed as either bar or line graphs.

Can you find examples of discrete and continuous variables in other investigations in this book?

Stretching to the Max!

You have investigated how the width of a rubber band affected its flight. What do you think would happen if you changed its length? That is, if you stretched the band different amounts. How might the band act when it hit the floor? Would it roll, slide, or not move? Follow the procedure for *Investigation 10.2, Stretching to the Max* to test your hypothesis.

Investigation 10.2

Stretching to the Max

What You Need

- 3 rubber bands of same width
- Metric ruler
- Flat surface such as a table

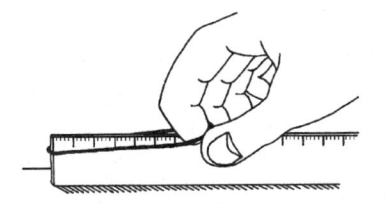

SAFETY: Do not shoot bands in the direction of people, animals, and objects. See Chapter 8, Experimenting Safely, Section E, Animals and Humans.

What You Do

1. Hold a ruler on its edge on a flat surface such as a table.
2. Place a medium rubber band so that one end of it just catches on the end of the metric ruler.
3. Pull the other end of the rubber band until the sides of the rubber band are flat against the ruler but the rubber band is not stretched. Release the rubber band.
4. Measure how far the rubber band moved before it hits the floor. Observe its behavior once it hits the floor. Use the following categories to describe behavior: No Movement (N), Slide (S), or Roll (R).
5. Repeat Steps 2-4 stretching the rubber band to 1, 2, and 3 cm beyond its length in Step 3.
6. Repeat Steps 2-5 for a total of 4 trials.

First, let's look at the data we collected on the effect of stretch on the distance the rubber band traveled (see Table 10.2).

TABLE 10.2 The Effect of Stretching a Rubber Band on Distance Traveled

Distance Stretched (cm)	Distance Traveled (cm) Trials				Mean Distance (cm)
	1	2	3	4	
0	5	2	5	4	4
1	105	58	114	76	88
2	141	190	160	171	165
3	278	254	160	310	252

Constructing Bar Graphs

To construct a bar graph of the data follow the same steps that you used to make a graph for the distance traveled by various width rubber bands. In summary, here are the steps.

Step 1. Draw and label the axes. Place the independent variable (distance stretched) on the horizontal (X) axis. Place the dependent variable (distance traveled) on the vertical (Y) axis. For the dependent variable you will use the mean distance. Remember to place your units of measurements, such as centimeters, in parenthesis.

FIGURE 10.2A

Step 2. Determine the scale for the horizontal (X) axis. Because you stretched the rubber band four different amounts, subdivide the axis into four parts and label each part (0, 1, 2, 3).

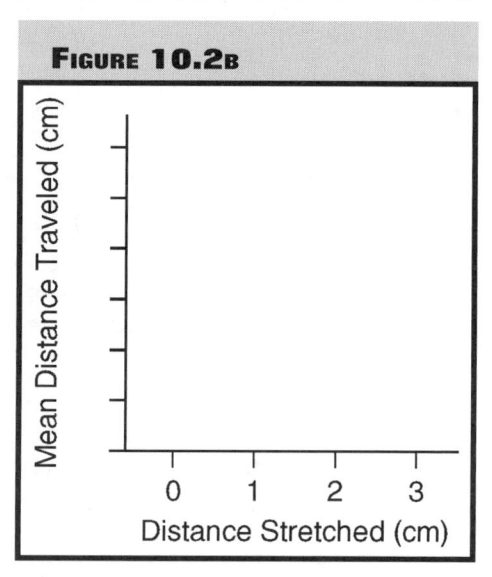

FIGURE 10.2B

Step 3. Determine the scale for the vertical (Y) axis. Find the difference between the largest and smallest numbers you will need to display. Be sure to use the mean distance. Divide the difference by 5 if you want about 5 intervals, and round the difference to an easy counting number.

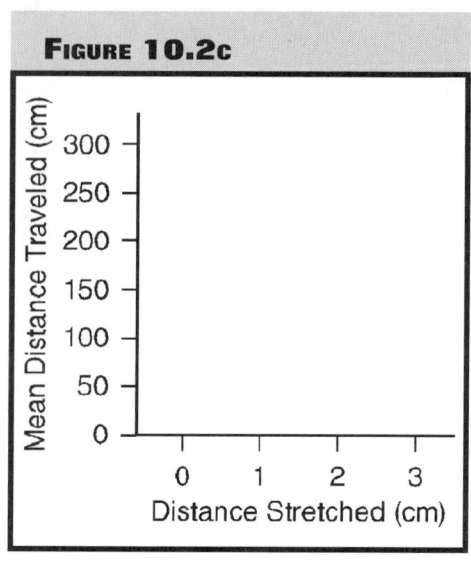

FIGURE 10.2C

Largest number	252
Smallest number	4
Difference	248
Difference divided by 5	248 ÷ 5 = 49.6
Quotient rounded to a counting number	50

To subdivide the scale, find a multiple of 50 below the smallest number (4) to be graphed and count upwards until the largest number (252) has been exceeded by a multiple of 50.

In this step you are asked to divide by 5 so that you will have a scale with approximately 5 to 7 intervals. If you have too few intervals on the scale, such as 2 or 3, you will have difficulty graphing the data. If you use too many intervals on your scale, such as 12 or 15, the scale will become crowded and difficult to read.

Step 4. Plot the data. Because the rubber band traveled 4 cm when it was stretched 0 cm, draw a vertical bar from the amount of stretch (0 cm) upwards until it reaches the appropriate amount of stretch (4 cm). Continue this way for stretches of 1, 2, and 3 cm until the graph is completed.

Step 5. Write about the pattern on the graph.

The more the rubber band was stretched, the greater the distance traveled across the floor.

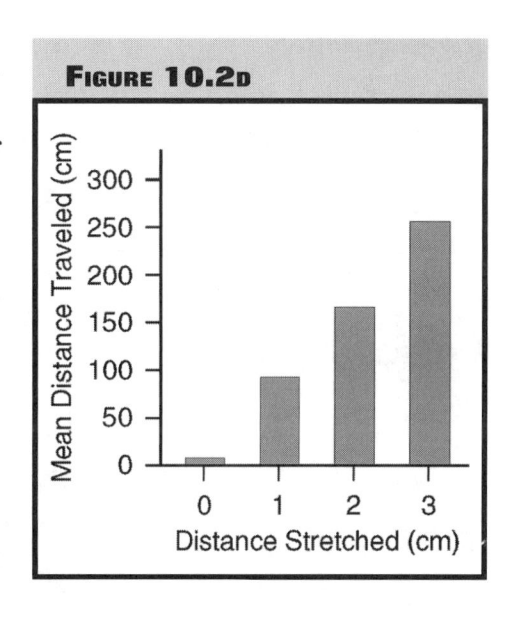

FIGURE 10.2D

Bars versus Lines

The distance that you stretched the rubber band was a continuous variable. Even though you stretched the band by whole numbers such as 1, 2, and 3 cm you could have used values such as 1.5 cm or 2.3 cm. Because the points between the values tested have meaning, continuous variables can be displayed as bar or line graphs. To make a line graph, rather than a bar graph, make a dot where the midpoint of the bar would be and remove the bars. Then, draw a line so that about half the dots fall to either side (see Figure 10.3).

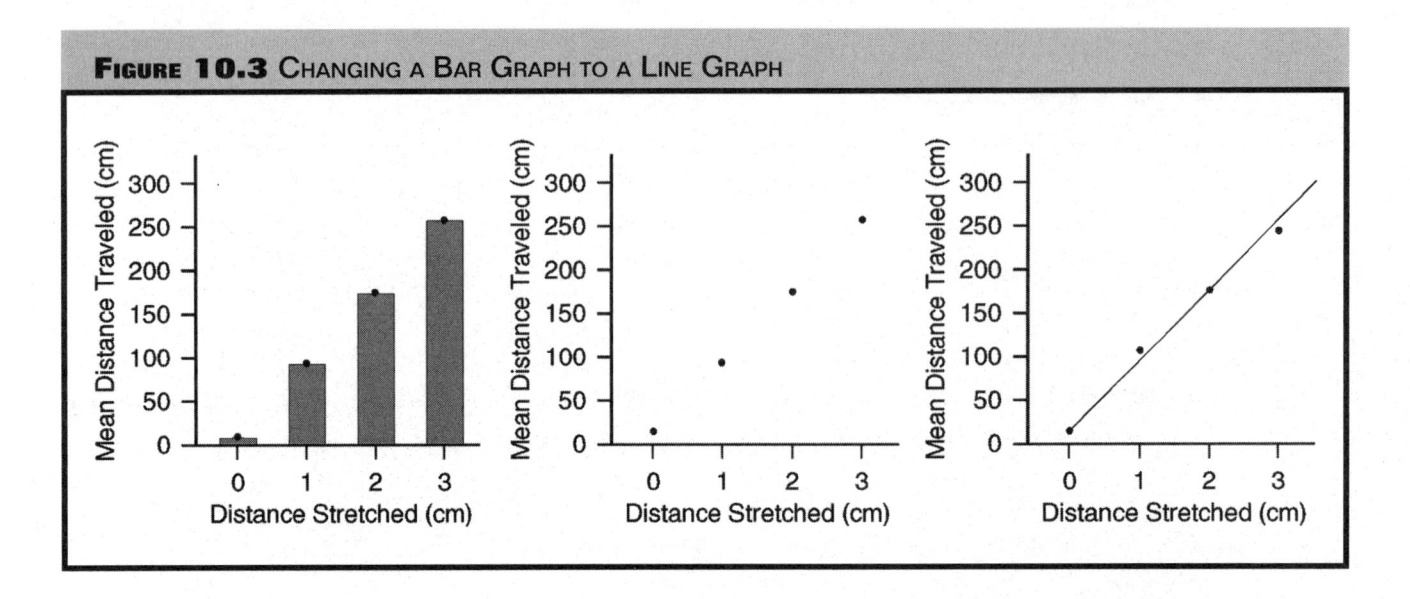

FIGURE 10.3 CHANGING A BAR GRAPH TO A LINE GRAPH

From the line graph, it is much easier to predict values. For example, if the distance stretched were 1.5 cm, how far do you predict the rubber band would travel? What about when the band was stretched 2.5 cm? With a line graph, it is also easier to see the pattern or relationship between the variables. In Chapter 11 you will learn how to construct a line graph without having to make a bar graph first.

Rolling Along

When your rubber band hit the floor, how did it behave? Did it not move? slide? or roll? See the data we collected in Table 10.3.

The independent variable, amount of stretch, is continuous so you have a value to graph on the horizontal (X) axis. But what do you put on the vertical (Y) axis? You cannot graph the modes—slide, roll, or no movement. Instead, you show the variation in your data by graphing the number of times each category occurred. Bar graphs that show the number of times something happened are called **frequency distributions**. To make a frequency distribution, just follow the same steps as for other bar graphs.

TABLE 10.3 The Effect of Stretching a Rubber Band on Landing Behavior

Amount of Stretch (cm)	Type of Behavior Trials 1	2	3	4	Mode	Frequency Distribution
0	N	S	N	N	N	N: 3 S: 1 R: 0
1	S	S	R	S	S	N: 0 S: 3 R: 1
2	S	R	R	S	S-R[1]	N: 0 S: 2 R: 2
3	S	R	R	R	R	N: 0 S: 1 R: 3

Key: N = No movement   S = Slide   R = Roll

[1] Sometimes there are two or more modes in a set of data.

Step 1. Draw and label the axes. Place the independent variable, distance stretched, on the horizontal (X) axis. Label the vertical (Y) axis—number of rubber bands.

FIGURE 10.4A

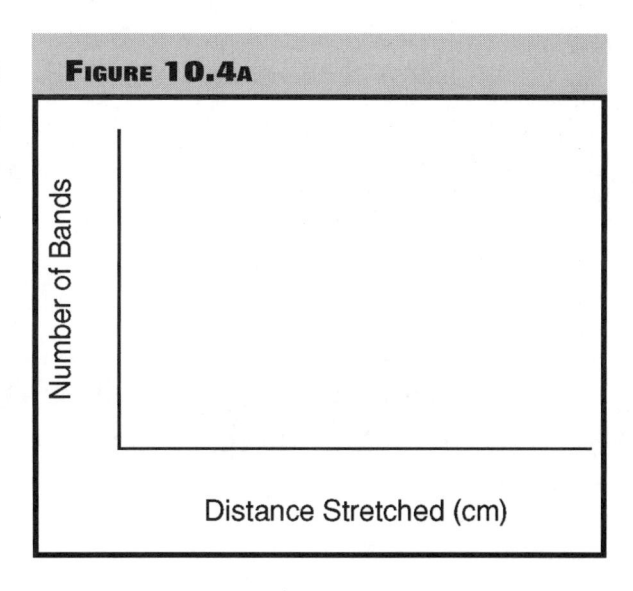

Y-axis: Number of Bands
X-axis: Distance Stretched (cm)

Constructing Bar Graphs   117

Step 2. Determine the scale for the horizontal (X) axis. First, subdivide the axis into four areas to represent the different amounts you stretched the band—0, 1, 2, 3 cm. Then, subdivide each of these areas for the three categories of behavior—No Movement (N), Slide (S) and Roll (R).

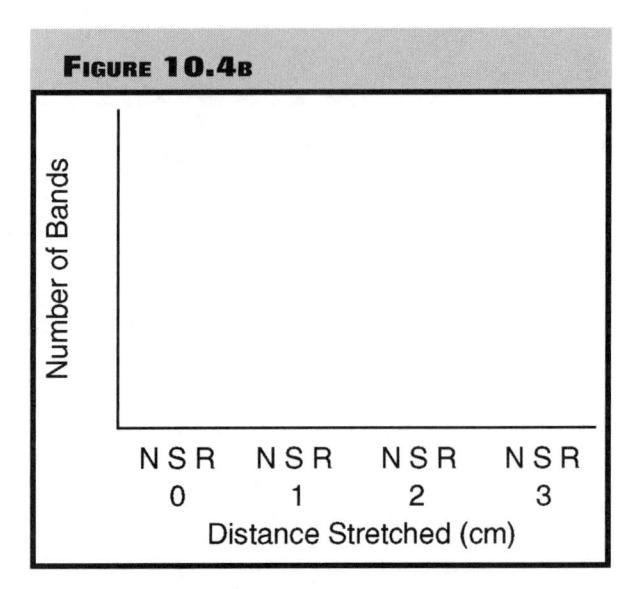

FIGURE 10.4B

Step 3. Determine the scale for the vertical (Y) axis. From the data table, determine the smallest (0) and the largest (3) number of occurrences of landing behaviors you will need to graph. Decide on an appropriate interval for the scale.

FIGURE 10.4c

Largest number	3
Smallest number	0
Difference	3
Difference divided by 5	0.6
Quotient rounded to a counting number	1

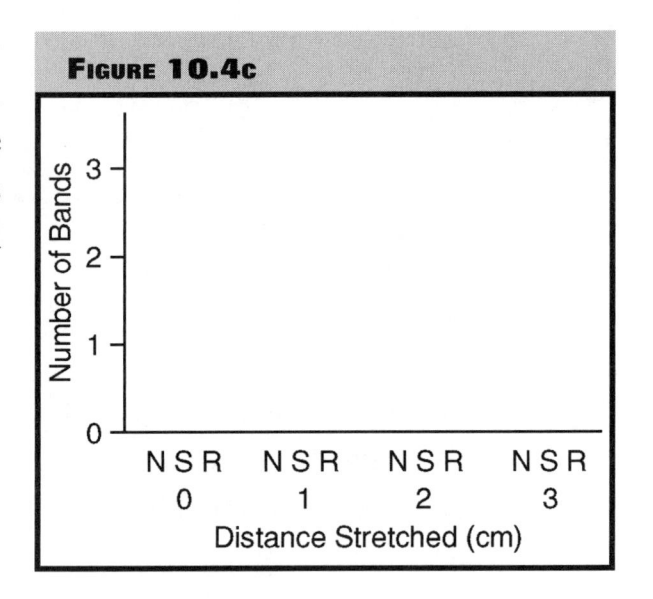

With frequency distributions, you will always round to a whole number. To subdivide the scale, simply count upwards from 0 to 3 by ones.

118   Chapter 10

Step 4. Plot the data. First, determine a symbol, such as solid, striped, and dotted bars that you will use to represent each category of rubber band behavior. In the upper right hand corner of your graph place a key to describe the symbols. Then, begin plotting your data.

FIGURE 10.4c

Look at the stretch distance of 0 cm and note the number of no movement (3), sliding (1), and rolling (0) rubber bands. In the area assigned to 0 cm, draw a vertical bar from the behavior category, no movement, to the corresponding number of occurrences (3). Similarly, draw a bar for the number of sliding (1) occurrences. When there are no occurrences (0), leave a blank space on the graph.

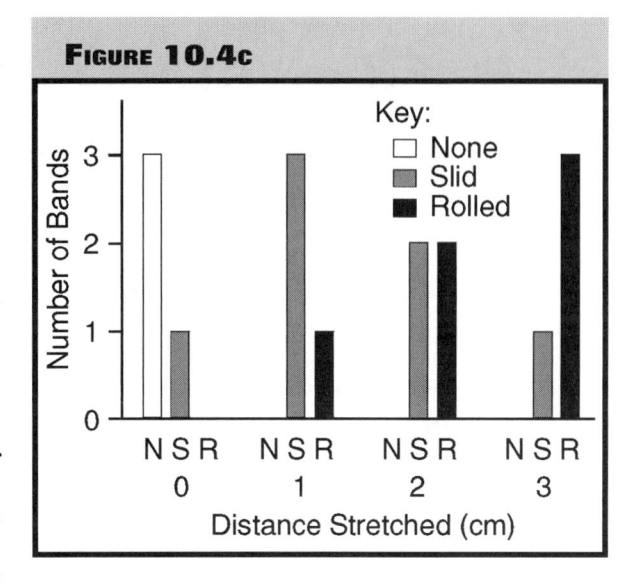

Using the same procedure, complete the frequency distribution for stretches of 1, 2, and 3 cm.

Step 5. Write a few sentences about the data:

> When rubber bands are stretched, they tend to slide or roll on a flat surface. The more the rubber band is stretched, the more likely it is to roll.

Applying Your Skills

Use your skills to make bar graphs for some of the practice problems. Then, apply your skills to real data by conducting *Investigation 10.3, Inflatable Foods?*

Constructing Bar Graphs 119

Investigation 10.3

Inflatable Foods?

What You Need

- 4 small zip bags (sandwich or smaller)[1]
- 5 mL metric measuring spoon
- Graduated cylinder (50 mL)
- 6 packages of dry baker's yeast
- Sugar
- Flour
- Cookies such as animal crackers
- Warm water (40° to 50° C)
- Marker or tape and pencil
- Metric ruler
- Sheet of paper
- Safety goggles

[1]If you use smaller snack-size ziplock bags, you can reduce the amount of yeast and water used. Use 2 mL yeast to 60 mL warm water.

See Chapter 8, Experimenting Safely, Section A, Chemicals.

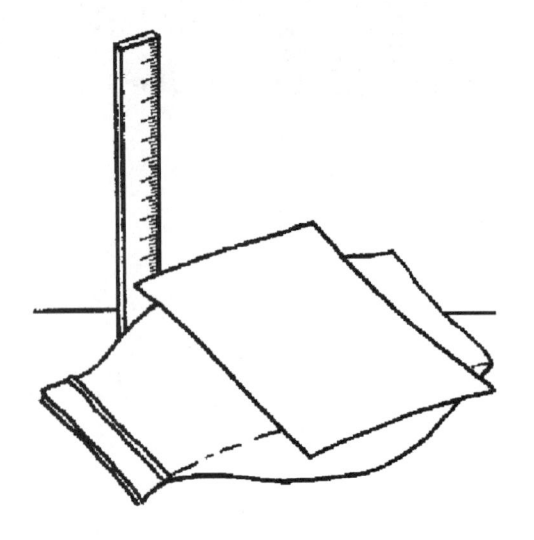

What You Do

1. Label each of the bags—yeast, flour, sugar, and cookies.
2. Place 5 mL[1] of yeast in each labeled bag.
3. Add 125 mL[1] of warm water to each bag. The water should be approximately 40° to 50° C. This is equivalent to warm water from the faucet. Do not use boiling or very hot water. It will kill the yeast.
4. Add 10 mL of flour, sugar, and cookies to each of the labeled bags.
5. Be sure all the air is removed by zipping the bag almost all the way. Leave an opening about the size of your finger. Lay the bag flat on a surface while holding the open corner slightly elevated. Press the bag to force the air out. Then, finish *sealing* the opening.
6. Fill a basin or sink with warm water. Place the bags in the basin or sink.
7. Watch the bags for evidence of a reaction. After 30 min. measure the height of the bag in millimeters and record the data. When measuring, you may find it helpful to lay an index card across the bag and line it up with your ruler.
8. Describe the amount of puff as Not Rounded (NR), Slightly Rounded (SR) or Very Rounded (VR) and record your data.
9. Repeat Steps 1-7 for a total of 3 more trials.

Chapter 10

Analyzing Your Data

1. Make a data table to record the height (mm) that the bag puffed. Find the average height and range for yeast, flour, sugar, and cookies. Construct a graph to display your data. Write about the data.

2. Make a data table to record the amount (NR, SR, VR) that the bags puffed. Find the mode for yeast, flour, sugar, and cookies. Construct a frequency distribution to display your data. Write about the data.

For more information on Constructing Bar Graphs, use these search words: or phrases

"bar graphs"
axes
histograms

to find websites such as this:
exploratorium.edu/learning_studio/ozone/graphing1.html

Practice Problems

Constructing Bar Graphs

1. Construct a bar graph to display the data. Write sentences to summarize the trends.

A. Title: The Effect of Oil Brand on Time for Drops of Water to Fall Through the Oil

Brand of Oil	Time for Water Drop to Fall (sec) Trials			Mean Time (sec)	Range (sec)
	1	2	3		
A	10	15	12	12	5
B	45	38	48	44	10
C	15	19	22	19	7
D	65	78	64	69	14

B. Title: The Effect of Kind of Fruit on the Time It Takes to Turn Brown

Kind of Fruit	Time to Turn Brown (min.) Trials				Mean (min.)	Range (min.)
	1	2	3	4		
Apple	40	38	45	42	41	7
Pear	25	30	34	28	29	9
Peach	18	20	22	19	20	4

> ## Remember
>
> The **independent variable** is on the horizontal line, or X axis, of the graph.
>
> The **dependent variable** is on the vertical line, or Y axis, of the graph.

2. Construct a bar graph for the data and write sentences to summarize the trends.

A. The Effect of Salad Oil Temperature on the Time for a Water Drop to Fall Through It

Temperature of Salad Oil (°C)	Time to Fall (sec) Trials 1	2	3	Mean (sec)	Range (sec)
10	150	120	135	135	30
20	100	98	110	103	12
30	80	91	87	86	11
40	60	55	63	59	8
50	40	38	47	42	9

B. The Effect of Concentration of Antifreeze on Time for Water to Freeze

Amount of Antifreeze (%)	Time to Freeze (min.) Trials 1	2	3	4	Mean Time (min.)	Range (min.)
0	60	72	58	65	64	14
3	75	80	72	84	78	12
6	100	99	105	104	102	6
9	150	148	162	157	154	14

Constructing Bar Graphs

3. Construct frequency distributions for the data. Write sentences to summarize the results.

A. Title: The Effect of the Number of Drops of Blue Food Coloring on the Coloration of White Carnations

Amount of Food Coloring (Drops)	Coloration of Flower Trials					Mode	Frequency Distribution
	1	2	3	4	5		
2	S	S	S	S	S	S	S: 5 M: 0 L: 0
4	S	M	S	M	S	S	S: 3 M: 2 L: 0
6	M	M	M	L	M	M	S: 0 M: 4 L: 1
8	M	M	L	L	L	L	S: 0 M: 2 L: 3

Key: S = Small   M = Medium   L = Large

B. Title: The Effect of Rain Fall on the Cloudiness of Well Water

Amount of Rainfall (cm)	Cloudiness of Well Water Trials						Mode	Frequency Distribution
	1	2	3	4	5	6		
0.0	NC	NC	NC	NC	NC	NC	NC	C: 0 NC: 6
0.5	NC	NC	C	NC	NC	NC	NC	C: 1 NC: 5
1.0	NC	NC	NC	C	C	NC	NC	C: 2 NC: 4
1.5	NC	NC	C	C	C	NC	C-NC	C: 3 NC: 3
2.0	NC	C	C	C	C	C	C	C: 5 NC: 1

Key: NC = Not Cloudy   C = Cloudy

Constructing Line Graphs

Plots, Lines, and Trends

In Chapter 10 you learned to construct bar graphs, including a special kind of bar graph called a frequency distribution. You also learned that some data can be displayed as another kind of graph called a **line graph**.

The kind of graph that you can use to display data depends on the type of data. The data for some experiments can be displayed as either a bar graph or a line graph if both of the variables are continuous. Continuous means that the values of a variable are not separate categories and the intervals between the values have meaning. For example, suppose you dropped a ball from different heights, 70, 80, 90 and 100 cm, to see how drop height affects bounce height. Because drop height is a continuous variable, you could also drop the ball from other heights, such as 72 or 75. The interval between 70 and 80 has meaning. Other examples of **continuous variables** are amounts of water, weight of apples, and number of flowers. When the variables in an investigation are continuous variables, you can display the data as either a bar graph or a line graph.

> **Remember**
>
> **Continuous variables** are not separate categories, and the intervals between the values have meaning.
>
> **Discrete variables** are separate categories, and the intervals between them have no meaning.

When a variable consists of **discrete categories**, such as the kind of paper—newsprint, paper towel, and notebook, you must display the data as a bar graph, not a line graph. Remember, **discrete** means that the categories are separate and not continuous. The spaces or intervals between categories are not equal and thus, have no meaning; there is no kind of paper that is halfway between newsprint and paper towels. Other examples of discrete data are days of the week, types of skates, brands of batteries, and kinds of cars.

Try *Investigation 11.1, Collision Course*. Think about the variables you are investigating. What kinds of data are involved?

Investigation 11.1

Collision Course

WHAT YOU NEED

- 1 plastic ruler with a groove down its length
- Another ruler with Metric units
- 4 books or other objects to raise the ruler
- 1 marble or other small ball
- 1 plastic, foam, or paper cup
- Safety goggles

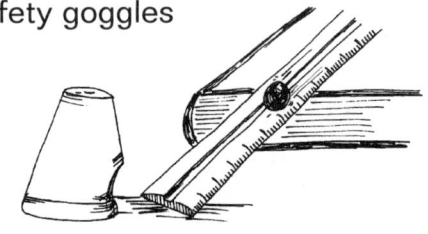

See Chapter 8, **Experimenting Safely.**

WHAT YOU DO

1. Construct a ramp using one book and the ruler with the center groove.
2. Cut away a section of a cup so the marble can easily roll into it when the cup is placed upside down.
3. Place the cup upside down in front of the ruler/ramp so the marble can roll into the opening in the cup.
4. Measure the height of the ramp (at the highest point), and then roll a marble down the ramp.
5. Measure how far the cup moved after the marble crashed into it. Record this distance.
6. Place the cup back in its original position and roll the marble down the ramp at this height 2 more times for a total of 3 repeated trials.
7. Repeat Steps 3-6, changing the height of ramp by using 2, 3, and 4 books.

Your data:

Height of the Ramp (cm)	Distance Cup Moved (cm) Trials			Mean Distance Cup Moved (cm)	Range (cm)
	1	2	3		

126   Chapter 11

Our data:

TABLE 11.1 The Effect of Height of Ramp on the Distance a Cup Moved

Height of the Ramp (cm)	Distance Cup Trials 1	2	3	Mean Distance Cup Moved (cm)	Range (cm)
5	4	4	5	4	1
7	8	9	8	8	1
12	15	15	16	15	1
14	21	22	20	21	2

How does your data compare with ours? Your data may be different because some of your constants—the size of the cup, the mass of the marble, and the kind of surface—are different from ours.

Selecting the Right Kind of Graph

What kind of graph, a bar or a line graph, should you use to display the data you collected in *Investigation 11.1, Collision Course*? Determining the right type of graph to use can be as easy as answering two questions about your independent and dependent variables:

1. *Is the independent variable discrete or continuous?*

    If the IV is discrete, you have no choice. You must do a bar graph; you can't make a line graph.

    If the IV is continuous, then you must also look at your dependent variable before deciding on a bar or line graph.

2. *Is the dependent variable quantitative or qualitative data?*

    If the DV is qualitative (data that are collected by observing, describing, and placing things in categories), then your only option is a bar graph. In fact, you must use a special kind of bar graph called a frequency distribution (see Chapter 10).

    If the DV is quantitative (data that are counted or collected using a standard measuring scale), then you have two options—line or bar graphs.

Use Table 11.2, **Choosing the Appropriate Graph**, to help you determine the right graph for Investigation 11.1.

TABLE 11.2 Choosing the Appropriate Graph		
Type of Independent Variable	Type of Dependent Variable	Appropriate Types of Graphs
Discrete	Quantitative Qualitative	Bar Graph Frequency Distribution
Continuous	Quantitative Qualitative	Bar or Line Graph Frequency Distribution

Constructing Line Graphs

Let's use the data in Table 11.1 to construct a line graph such as the one shown in Figure 11.1. To construct a line graph, follow the same steps you learned in Chapter 10.

Step 1. Draw and label the axes. Place the independent variable (height of ramp) on the horizontal (X) axis. Place the dependent variable (the mean or average distance the cup moved) on the vertical (Y) axis. Remember to include the units of measurements, such as centimeters in parenthesis (cm).

FIGURE 11.1 THE EFFECT OF HEIGHT OF RAMP ON THE DISTANCE A CUP MOVED

Step 2. Determine the scale to use for the horizontal axis by first finding the range of the data to be graphed. Subtract the smallest value (5) from the largest value (14) of the independent variable in Table 11.1, 14 − 5 = 9. Divide this difference by the number of intervals you want. If you want about 5 intervals, divide by 5. This

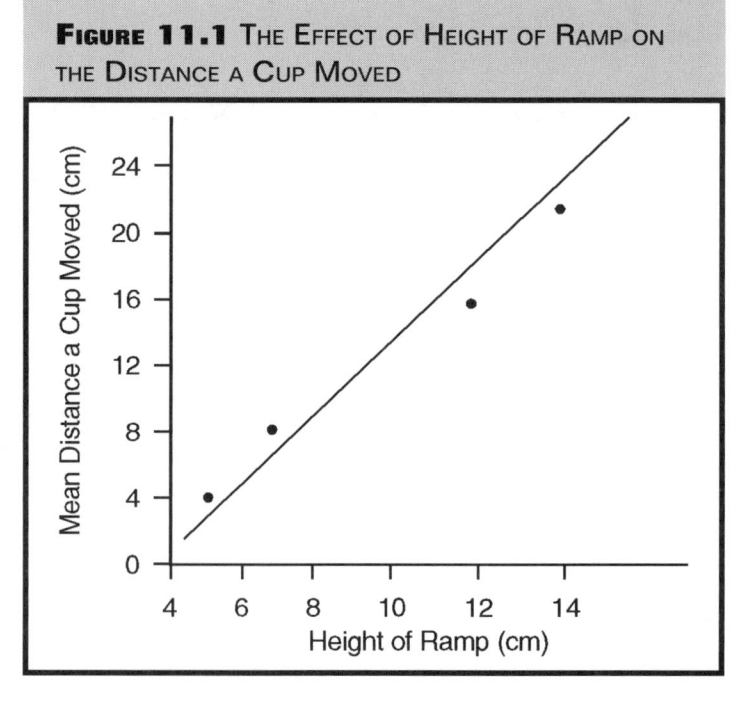

128 Chapter 11

usually results in a scale with 5 to 7 intervals. Too many intervals crowds a graph, while too few makes it difficult to plot points. Dividing 9 by 5 = 1.8

Using 1.8 to make intervals would be too difficult, so make the job easier by rounding 1.8 to an easy counting number like 2. Other good counting numbers are multiples of 5, such as 10 or 20, or other small numbers such as 4. Use this rounded number to mark off the intervals. Begin with a multiple of 2 that is less than the smallest value (5) to be plotted and continue until you have reached or exceeded the largest value (14).

Step 3 Repeat this procedure using the mean distance data to determine the scale for the vertical axis:

Largest value	21
Smallest value	4
Difference	17
Difference divided by 5	17 ÷ 5 = 3.4
Quotient rounded to a counting number	4

To label the vertical axis, begin with a multiple of 4 below the smallest number to be graphed. In this case the smallest number is 4 so it is appropriate to begin the scale with 0.[1] Then count upwards with multiples of 4 until you have reached a number larger than the largest number (21) to be graphed.

Step 4 Plot the data. Refer to Table 11.1. The first data pair to be plotted is (5,4). When writing data pairs, the first number in the pair is for the independent variable, while the second number is for the dependent variable. In the example of (5,4), the 5 is the first level of the IV that was tested and the 4 was the average result (dependent variable) of the three

---

[1] Some people feel that all graphs should begin with the intersection of the X and Y axes labeled (0,0). This works fine when the data sets start at or near zero. But in many experiments, data sets begin with numbers far from zero, resulting in a large gap between 0 and the first piece of data to be graphed. If you or your teacher are uneasy about not beginning at zero, you can use the symbol (⨍) used in Figure 10.1 to indicate that part of the graph is not shown, such as the intervals below 520 cm in the example in Chapter 10 on page 111.

trials. In Figure 11.2, locate 5 on the horizontal axis and 4 on the vertical axis. Imagine a **vertical** line drawn straight up from the 5 and a **horizontal** line drawn straight across from the 4. Where these two imaginary lines meet is a point representing that data pair.

FIGURE 11.2 THE EFFECT OF HEIGHT OF RAMP ON THE DISTANCE A CUP MOVED

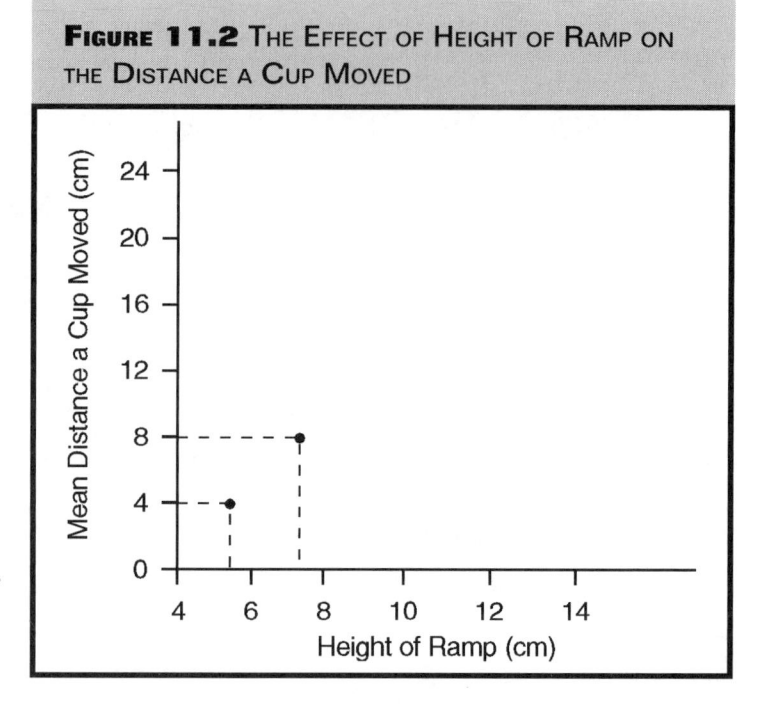

Look at the second pair of data in the table (7,8). Sight imaginary lines straight up from the 7 and straight across from 8. The point at which these imaginary lines intersect is a point representing this data pair.

Step 5 Look for patterns in the data points.

Looking for Patterns

After plotting all the points for the data pairs, look for a pattern in the points. Is there a general upward trend of the points? Is there a trend downward? Do the points go up and then level off? Or do the points gradually increase, reach a peak, and then gradually decrease?

FIGURE 11.3 EXAMPLES OF LINES-OF-BEST-FIT

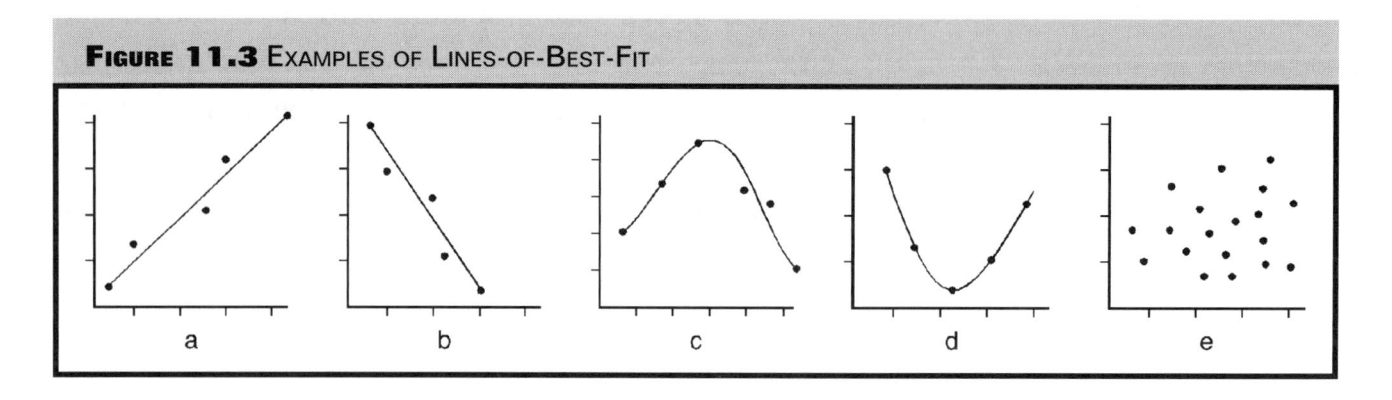

Notice in Figure 11.3a that the data points tend to move up as you look from left to right on the graph. This means that as the independent variable *increases*, the

130   Chapter 11

dependent variable also *increases*. The pattern in Figure 11.3a is called a **positive association**. In other graphs, such as 11.3b, the reverse is true. The data points tend to move down as you look from left to right on the graph. This means that as the independent variable *increases*, the dependent variable *decreases*. The pattern in Figure 11.3b is called a **negative association**. Other patterns of data are shown in Figures 11.3 c and d. These graphs indicate both positive and negative associations in the same graph. In Figure 11.3c the graph begins with a positive association between the variables, but then becomes a negative association. In Figure 11.3d the opposite is shown; there is first a negative association between the variables that changes to a positive association.

> **Remember**
>
> Data points that go up is a **positive association**.
>
> Data points that go down is a **negative association**.
>
> If there does not appear to be a pattern it is called **no association**.

Sometimes there does not appear to be any pattern in the points on a graph. When no pattern can be seen, this is called **no association** (see Figure 11.3e).

Lines of Best Fit

After you have identified a pattern or trend in the data points, try drawing a line that best shows that trend. Do not make a zig-zag line connecting the points. Draw the line so that about half the points are on one side of the line and half are on the other. Some points may actually be on the line (see Figure 11.1 on page 126). You can draw a better line-of-best-fit by collecting and plotting more points. Examples of lines-of-best-fit are shown in Figure 11.3.

Summarizing Trends

Graphs display data in pictorial form. You can turn that picture into words by writing a sentence or two that summarizes the general trend of the data as shown by the line-of-best-fit. When writing a sentence, describe the effect of changing the independent variable on the dependent variable. For example, you could say, "As the height of the ruler ramp increased, the distance the cup moved also increased." Be sure you describe the general trend on the graph rather than write sentences that just repeat the data points.

Constructing Line Graphs

Applying Your Skills

Apply your skills to real data by conducting *Investigation 11.2, Curds and Whey*. Then, use your skills to make line graphs for some of the practice problems.

Investigation 11.2

Curds and Whey

WHAT YOU NEED

- Milk at various temperatures
- Thermometer
- 5 small clear glasses or beakers
- Watch or clock with a second hand
- Vinegar
- Metric or English measuring spoons
- Safety goggles

See Chapter 8, Experimenting Safely.

WHAT YOU DO

1. Put 2 mL of vinegar into 60 mL of ice cold milk 5°C.
2. Time the number of seconds it takes for the milk to completely solidify and form curds. Construct a data table and record your data. Record other observations such as curd color, curd size, curd texture, and color and amount of the liquid called whey.
3. To conduct additional trials, repeat Steps 1 and 2 three more times
4. Repeat Steps 1-3 at 15°, 25°, 35°, and 45°C. Record your data and observations.

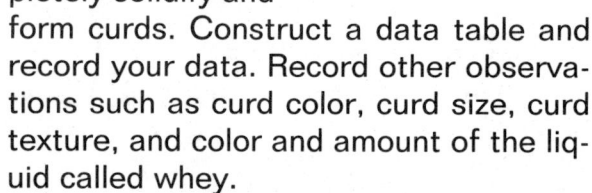

Analyzing Your Data

Use the time data in your data table to find the mean and range. Construct a graph to display your data and write a sentence or two that summarizes your data. If you recorded other observations such as curd color, size, and texture, what kinds of graphs would you use to display these data?

Searching the Web

For more information on Constructing Line Graphs use these search words or phrases:

"line graphs"
"linear relationships"
"data trends"
"plotting data"
"best fit lines"

to find websites such as this:
mste.uiuc.edu/courses/ci330ms/youtsey/intro.html

Practice Problems

Constructing Line Graphs

1. What kind of graph, bar or line, would you construct for this table of data?

Kind of Liquid	Evaporation Time (min.) Trials				Mean Evaporation Time (min.)	Range (min.)
	1	2	3	4		
Fresh water						
Salt water						
Alcohol						

2. What kind of graph, bar or line, would you construct for this table of data?

Height of Holes in Carton (cm)	Distance Water Spurts (cm) Trials				Mean Spurt Distance (cm)	Range (cm)
	1	2	3	4		

3. Write a sentence or two to describe the data trend in the following graph.

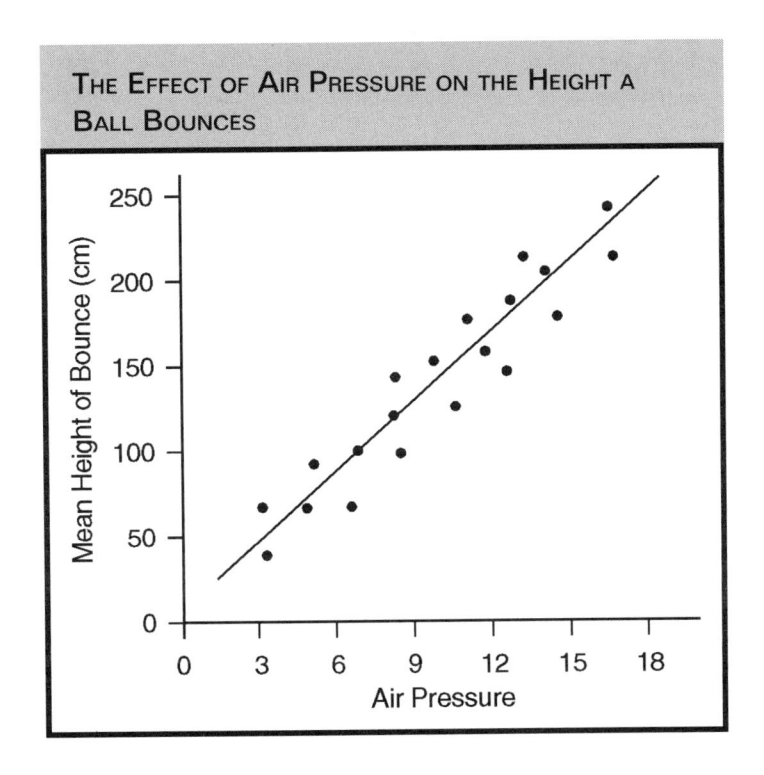

THE EFFECT OF AIR PRESSURE ON THE HEIGHT A BALL BOUNCES

(Scatter plot with line of best fit: x-axis "Air Pressure" from 0 to 18; y-axis "Mean Height of Bounce (cm)" from 0 to 250.)

4. Construct a line graph to display the following data. Write a sentence or two to summarize the trend of the data.

The Effect of Temperature of Water on Plant Height						
Temperature of Water (°C)	Plant Height (cm) Trials				Mean Plant Height (cm)	Range (cm)
	1	2	3	4		
25	33	34	32	32	33	2
30	30	28	28	31	29	3
35	24	23	21	23	23	3
40	18	16	17	14	16	4
45	11	12	10	12	12	2
50	8	12	11	11	11	4

Constructing Line Graphs

Notes

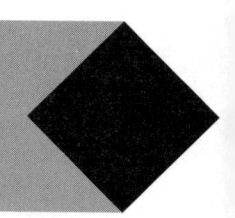

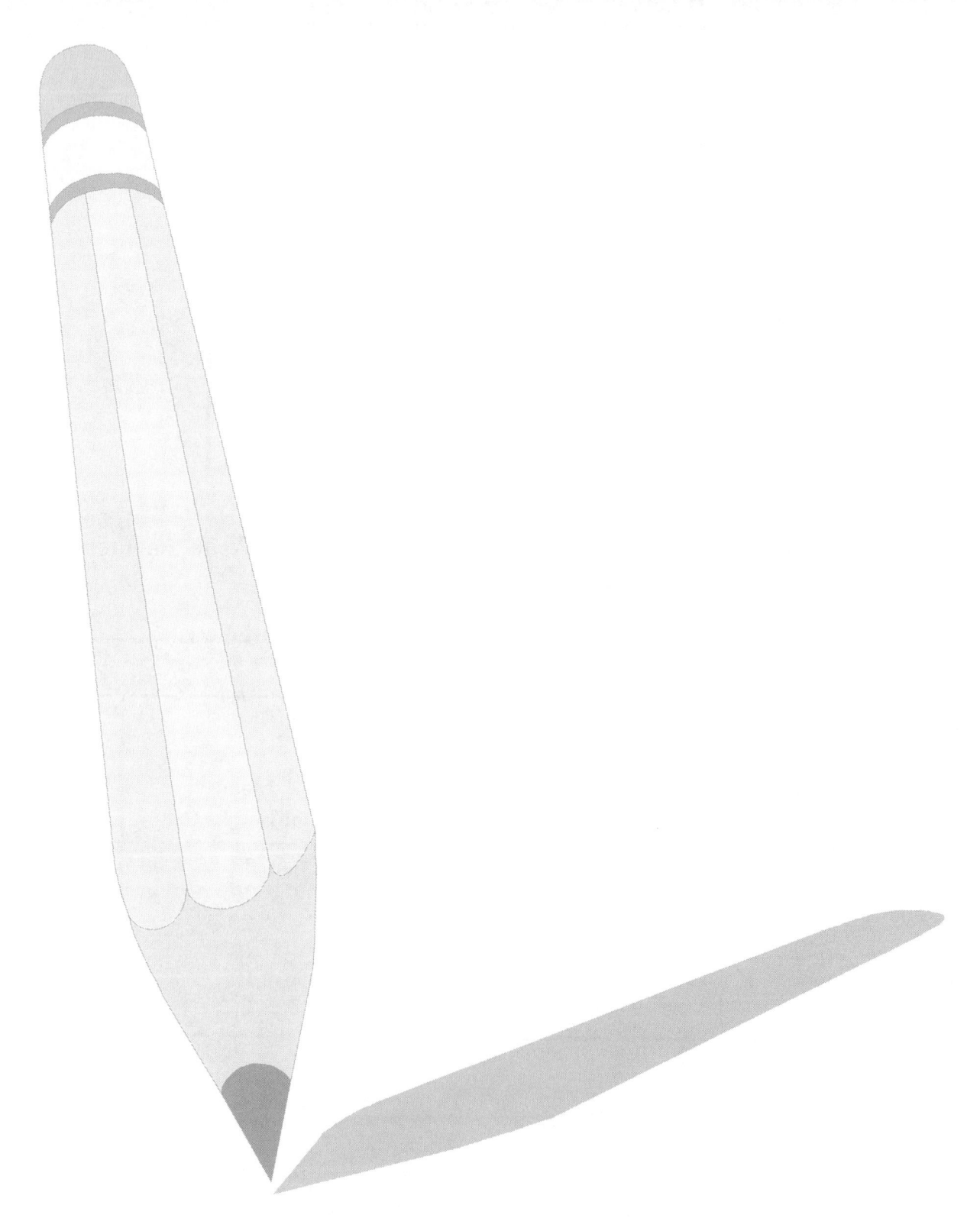

Writing an Experimental Report

From Introduction to Conclusion

Now it's time to write a report about your experiment. But, first let's answer **THE** important question: Why do scientists bother to write a report about their experiments? Is it necessary for you to write a report about your experiment? You did the experiment. You tested your hypothesis. You've learned what you wanted to learn. Why not get on with something else?

The answer is simple and straightforward. Experimental results must be shared. Each shared experiment adds to our knowledge about the universe. Unreported experiments do not add to this knowledge because only the investigator knows about them. No one else can learn from unreported results. Unreported experiments are so useless that scientists say, "The unreported experiment is an undone experiment."

Major Components

Of the seven parts of an experimental report, you have already learned how to do five of them. The seven major components of a simple report are:

Title	Learned in Chapter 3
Introduction	**WILL LEARN IN THIS CHAPTER**
Experimental Design Diagram	Learned in Chapter 3
Procedure	Learned in Chapter 7
Results (Data Tables, Graphs)	Learned in Chapters 9, 10, 11
Conclusion	**WILL LEARN IN THIS CHAPTER**
Bibliography	Appendix A: Using Style Manuals

An Introduction

The introduction section of the report tells the reader what the research problem was all about. It states your **reason**, or why you decided to study the topic you investigated. It also states your **purpose**, or what you hoped to learn by doing the experiment. Finally, the introduction states your **hypothesis**. The introduction to an experimental report provides answers to three questions:

1. Why did you conduct the experiment? **(Reason)**
2. What did you hope to learn? **(Purpose)**
3. What did you think would happen? **(Hypothesis)**

A Conclusion

A conclusion is a summary of an experiment. Someone who reads only the conclusion section of your report should be able to understand what your experiment was about. The summary should give your results, describe what those findings mean, and suggest new questions that should be investigated. A good conclusion can be written by answering six questions.

1. What was the purpose of the experiment?
2. What were the major findings?
3. Was your hypothesis supported by the data?
4. How did your findings compare with other research or with information in the textbook?
5. What possible explanations can you give for the findings?
6. What recommendations do you have for further study and for improving the experiment?

Sample Report

Study the science report on "The Effect of Over-Crowding of Bush Bean Seeds on Plant Height," that follows. The report contains all the major components of a simple science report. The three questions for writing an introduction and the six questions for writing a conclusion are indicated in the margin. This will allow you to see how the report answers the questions for the introduction and the conclusion.

The report is written in the third person using complete sentences. Each sentence includes enough of the question so that you don't need the question to understand the answer. It is important that the answers begin by phrasing the questions as statements. For example, Question 3 of the introduction, "What did you think would happen?" might be written like this: The researcher hypothesized that if the distance between seeds is decreased, then the height of the plants will decrease.

Title

The Effect of Over-Crowding of Bush Bean Seeds on Plant Height

Introduction

Question 1: Reason

The directions written on many seed packets state how far apart the seeds in the packet should be planted, such as 4 cm. They warn about crowding. The directions, however, do not describe how the growth of plants that are overcrowded will be harmed. Knowledge of how close together seeds can be planted without harm is important when growing space is limited.

Question 2: Purpose

The purpose of this experiment was to determine the effect of decreasing the distance between bush bean seeds on the height of the plants.

Question 3: Hypothesis

The researcher hypothesized that if the distance between the seeds was decreased below the recommended distance on the seed packet (4 cm), then the height of the plants that grew from the seeds would also decrease.

FIGURE 12.1 EXPERIMENTAL DESIGN DIAGRAM

Title: The Effect of Overcrowding of Bush Bean Seeds on Plant Height
Hypothesis: If the distance between the seeds is decreased below the recommended distance of 4 cm, then the height of the plants will decrease.

IV: Distance Between Seeds (cm)				
0	0.5 cm	1.0 cm	2.0 cm (control)	4.0 cm
10 seeds (trials)	10 seeds (trials)	10 seeds (trials)	10 seeds (trials)	10 seeds (trials)

DV: Height of plants (cm)
C: Container—flower box, 50 cm x 15 cm x 15 cm
 Soil—Bob's Potting, 10 cm deep
 Light—Fluorescent shop light at 30 cm above plants with 2 40-watt light bulbs, 12 hr. light per day from 7 am to 7 pm
 Fertilizer—none
 Water—250 mL sprinkled on plants every third day
 Seeds—bush bean

Procedure

Five flower boxes measuring 50 cm × 15 cm × 15 cm were filled to a depth of 10 cm with Bob's Potting Soil. In the first box 10 bush bean seeds were planted 0 cm apart. Ten seeds were planted 0.5 cm apart in the second box. In the third box the 10 seeds were planted 1 cm apart. In the fourth box the 10 seeds were planted 2 cm apart. In the last box the 10 seeds were planted 4 cm apart. Every third day 250 mL of water were sprinkled evenly on the soil in each box. The boxes were placed 30 cm below a fluorescent shop light. Each light contained two 40 watt regular fluorescent bulbs. The lights were turned on for a 12 hr. period each day from 7 a.m. to 7 p.m. The room was kept at 24°C. The height of the plants was measured in centimeters at the end of 30 days.

Results

TABLE 12.1 Effect of Over-Crowding on the Height of Bush Bean Plants

Distance Between Seeds (cm)	\multicolumn{10}{c	}{Height of Plants (cm) Trials}	Mean Height (cm)	Range
	1	2	3	4	5	6	7	8	9	10		
0	4	5	5	3	4	4	5	6	4	0	4	6
0.5	6	8	5	7	7	6	8	5	4	6	6	3
1	15	14	13	12	16	12	11	13	12	14	13	4
2	17	18	16	15	15	17	16	16	15	15	16	3
4	18	17	17	18	16	15	19	16	18	17	17	3

FIGURE 12.2 THE EFFECT OF DISTANCE BETWEEN BUSH BEAN SEEDS ON PLANT HEIGHT

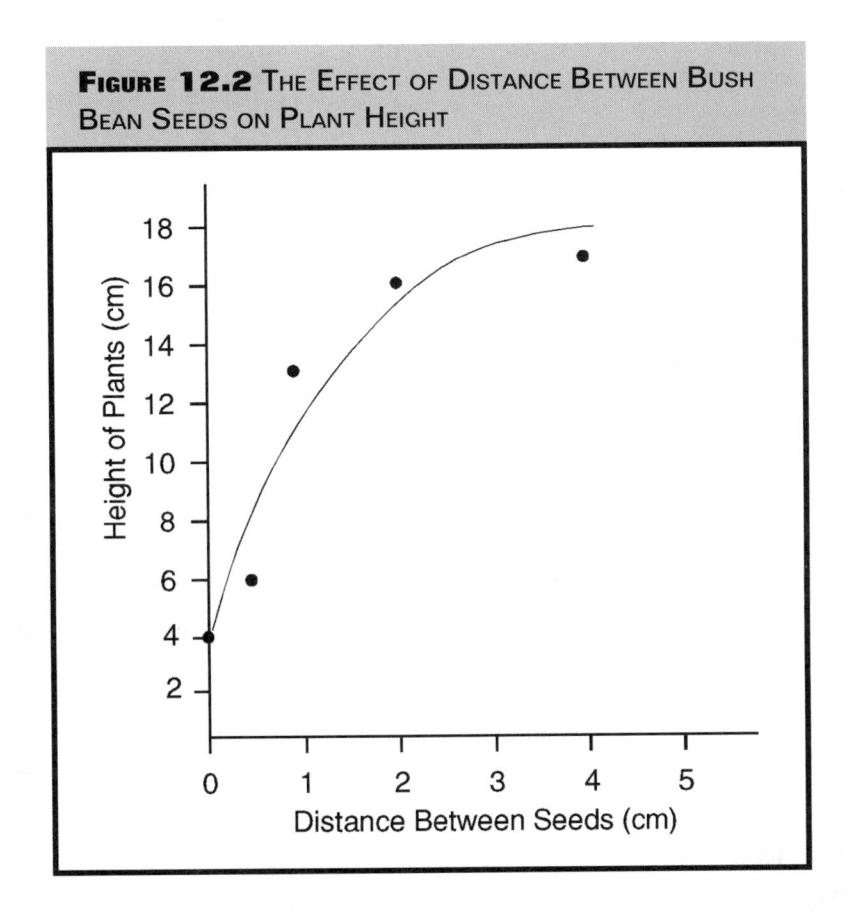

Summary Sentence: As the distance between the seeds increased, the height of the plants also increased.

Writing an Experimental Report 141

Conclusion

Question 1: *Purpose*

The purpose of this experiment was to determine the effect of decreasing the distance between planted bush bean seeds on the height that plants would grow.

Question 2: *Major Findings*

It was found that seeds planted close together were shorter. When seeds were planted very close, the increased distance made more of a difference. For example, when the distance was doubled from 0.5 to 1 cm the plant height increased by about 7 cm. However, when the distance was doubled from 2 to 4 cm the plant height only changed by about 1 cm. The ranges in plant heights were similar.

Question 3: *Support for Hypothesis*

The data supported the hypothesis that decreasing the distance between seeds would decrease the size of the plants.

Question 4: *Comparisons with Other Research*

These findings agree with the statements on seed packages and in plant books that warn to avoid overcrowding.

Question 5: *Explanation*

When the plants were planted closer together, they competed for the same light, air, water, and soil nutrients. Therefore, there were fewer materials for the plant to use in photosynthesis, the process by which it makes food. Without the food the plant could not grow.

Question 6: *Improvements and Other Recommendations*

This study should be repeated using other kinds of seeds. The effect of spacing distances between 2 and 4 cm should also be investigated. Possibly, plants could be grown slightly closer than 4 cm without causing much harm. The experiment could be improved by using more seeds and by measuring the plants more often.

Searching the Web

For more information on Writing an Experimental Report, use these search words or phrases:

"writing science reports"
"writing style manuals"
"writing conclusions"
"scientific writing style"

to find websites such as this:
kcs.kana.k12.wv.us/fair/researchpaper.htm

Practice Problems

Writing an Experimental Report

Now that you know the parts of a report, use the model to write an experimental report for a favorite experiment in this book or conduct *Investigation 12.1, Penny Barges* and use the data to write the report.

Investigation 12.1

Penny Barges

What You Need

- Plastic cup (approximately 250 mL or 8 oz.)
- Bowl or other container large enough for the plastic cup to fit on the bottom
- Metric or English measuring spoons or a tablespoon
- Table salt
- Water
- Paper towels, newspaper, etc.
- Pennies or other weights, such as washers
- Safety goggles

See Chapter 8, Experimenting Safely, Section A, Chemicals.

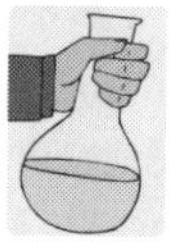

What You Do

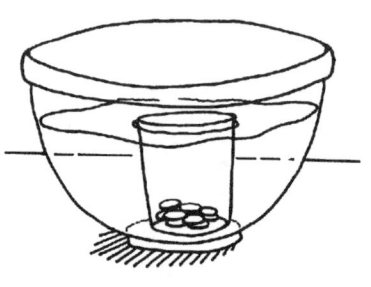

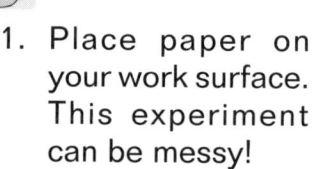

1. Place paper on your work surface. This experiment can be messy!

2. Hold the cup firmly on the bottom of the bowl and fill the bowl with water until the water rises almost to the rim of the cup.

3. Let the plastic cup float on the water. Add enough pennies to the cup so that it rests firmly on the bottom of the bowl. Count the number of pennies you used and record the data in a table. Remove the cup and repeat 2 more times for a total of 3 trials.

4. Totally dissolve 60 mL of salt in the water. Put the cup into the salt water. Add just enough pennies to make the cup rest firmly on the bottom again. Count the number of pennies you used and record the data. Remove the cup and repeat 2 more times for a total of 3 trials.

5. Repeat Step 4 using a total of 120 mL and 180 mL of salt.

144   Chapter 12

Communicating Your Findings

1. Construct an appropriate data table. Calculate means and ranges for the data.

2. Make a line graph to display your data. Summarize the trends on the graph.

3. Using the model in this chapter, write an experimental report of the experiment containing a title, introduction, experimental design diagram, procedure, results (data table and graph), conclusion, and bibliography.

4. Read Appendix C on expanding the introduction. Using the questions in Appendix C as a guide, write an expanded introduction for the report.

> **Remember**
>
> There are 7 components of an experimental report:
> 1. Title
> 2. Introduction
> 3. Experimental Design Diagram
> 4. Procedure
> 5. Results (Data Tables, Graphs)
> 6. Conclusion
> 7. Bibliography

Notes

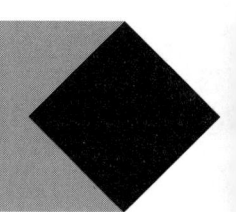

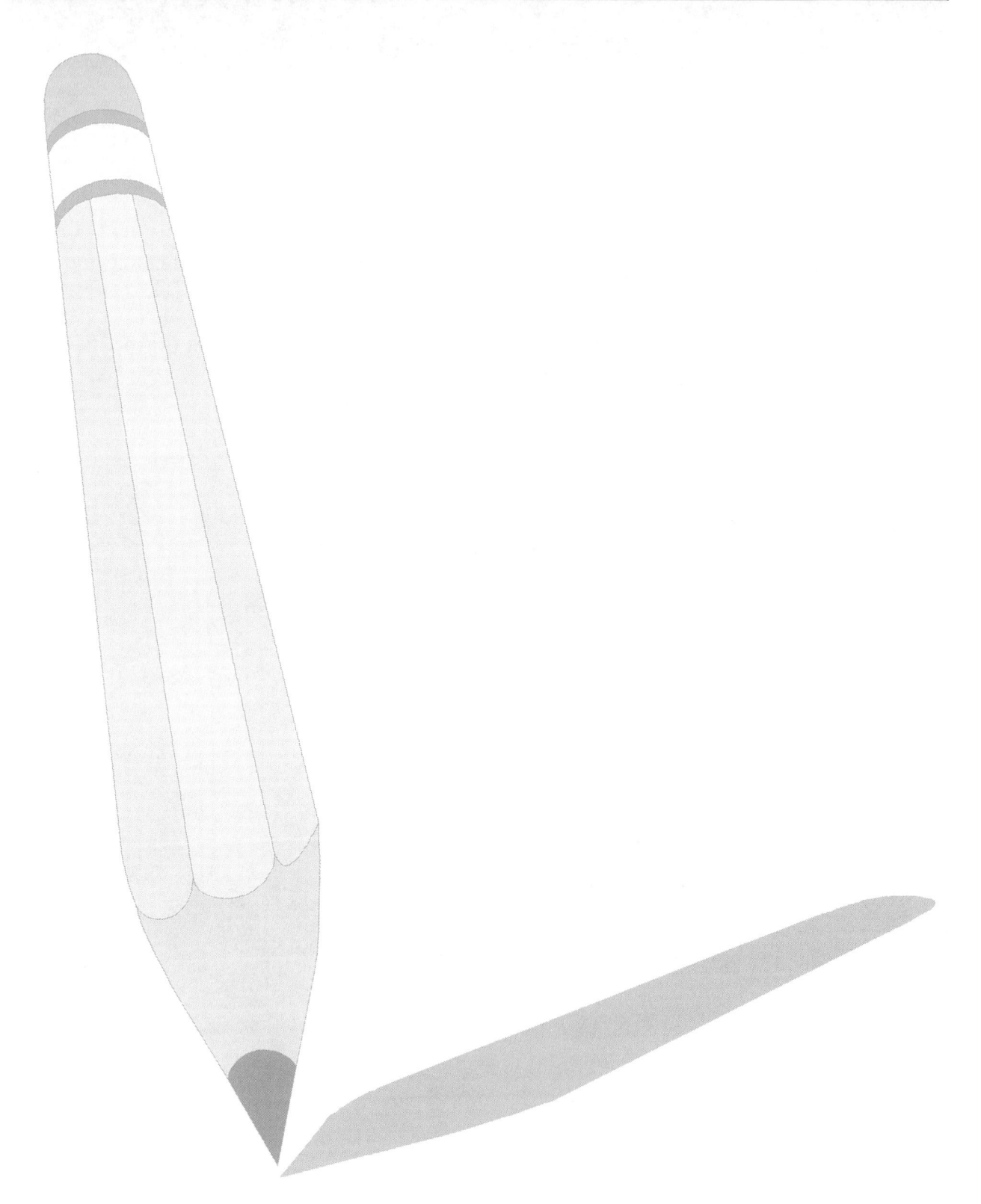

Making Stem-and-Leaf Plots

Data Displays and Distribution Patterns

Grab a handful of chips! Scoop up a handful of wrapped candies! Grabbing and scooping – both are ways to describe obtaining food, as well as other objects. Which is the really the best way? Which will result in picking up the largest quantity of objects? Predict whether scooping or grabbing would allow you to pick up the most objects and test your hypothesis using marbles, plastic bottle caps, or shell pasta pieces.

Investigation 13.1

Scooping or Grabbing?

WHAT YOU NEED

- Bowl or other container
- Approximately 1 liter of shell pasta
- Pan or other flat container with sides

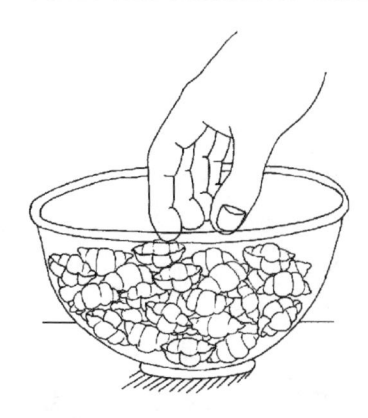

See Chapter 8, **Experimenting Safely.**

WHAT YOU DO

1. Fill the bowl with shell pasta.
2. Hold your dominant hand over the shell pasta with your fingers *pointed down*. For most people, the right hand is dominant, but if you are a "lefty" use your left hand.
3. Grab a handful of shell pasta and put it in the pan. Count the number of pasta pieces and record the information in a data table.
4. Return the pasta to the bowl.
5. Enter your data in a class data set. Repeat Steps 1 to 4 until at least 25 trials are recorded in the class data set.
6. Hold your dominant hand over the pasta with your fingers *pointed up*. Cup your hand and scoop up as many of the shell pasta pieces as possible. Record your data in a class data table. Repeat as needed so there are at least 25 trials recorded in the class data table.

Look at the data we gathered on medium-sized shell pasta. Can you see a pattern? Probably not, 31 trials are lots of data! To see if there is a pattern, you can apply a new technique called a stem-and-leaf plot.

TABLE 13.1 Effect of Technique on Number of Shell Pasta Obtained

Technique			
Grabbing		Scooping	
11	23	30	77
33	35	43	53
43	29	33	70
17	37	56	57
52	23	36	70
34	33	38	63
18	25	33	69
15	38	55	26
36	24	37	74
47	33	41	57
52	20	49	67
12	39	41	22
33	22	53	57
13	21	46	58
48	53	50	41
	16		47

Interpreting Stem-and-Leaf Plots

A stem-and-leaf plot is a type of graph in which each of the data points is plotted. It is a quick way to display 25 or more pieces of data. Each piece of data is displayed as two parts: a stem and a leaf. Before making a stem-and-leaf plot, let us first interpret some displays from an experiment involving brine shrimp or "sea-monkeys" conducted by a class of 26 students. Each student put 0.1g of brine shrimp eggs in a dish containing 30 mL of a 3% salt solution. After two days, each student used a magnifying glass to count the number of hatched brine shrimp. Figure 13.1 shows a stem-and-leaf plot of the brine shrimp data.

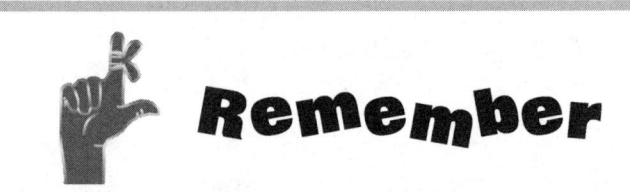

> **Remember**
>
> A stem-and-leaf plot is a quick way to display 25 or more pieces of data.

In a stem-and-leaf plot, a number such as 36 is displayed as |3|6 where 3 is the stem and 6 is the leaf. Every stem-and-leaf plot has a key that tells you how to interpret the graph. For example, the key |3|6 tells you that the stem is the tens digit and the leaves are the ones digit. Other examples are:

|0|6 = 6

|2|4 = 24

|11|5 = 115

|100|8 = 1008.

FIGURE 13.1 ORDERED STEM-AND-LEAF PLOT FOR NUMBER OF BRINE SHRIMP HATCHED IN A 3% SALT SOLUTION

Stem	Leaf
0	7 8
1	1 4 5 5
2	3 4 4 5 6 7
3	3 4 6 6 6 7 8 9
4	2 3 3 4
5	6 7

Key |3|6 = 36

Look at Figure 13.1 and answer the following questions about the data. Then, check your answers on page 149.

1. What values are represented by |1|5, |4|3, |2|6 ?

2. What was the minimum number of brine shrimp hatched?

3. What was the maximum number of brine shrimp hatched?

4. What was the range of the hatched brine shrimp?

5. What is the middle number (median) of brine shrimp hatched?

6. What was the mode of brine shrimp hatched?

7. Turn the stem-and-leaf plot so that the stem is on the bottom. Do the "leaves" make a pattern? If so, describe it.

Remember

To find the **range**, subtract the minimum value from the maximum value.

The **median** is the middle value in a set of data ranked lowest to highest.

The **mode** is the value that occurs most often.

Making Stem-and-Leaf Plots **149**

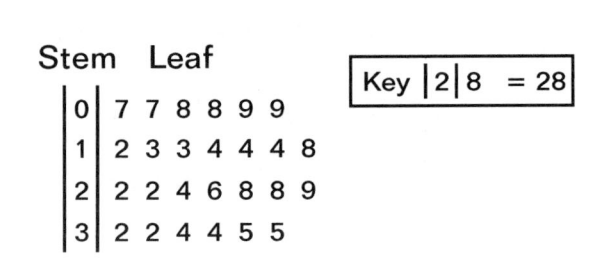

> **Answers to Questions about Figure 13.1**
>
> 1. 15, 43, 26
> 2. 7
> 3. 57
> 4. 57 – 7 = 50
> 5. 33.5
> 6. 36
> 7. Yes. Mound or bell-shaped with the number of "highs" balancing the number of "lows."

Now, let us look at another stem-and-leaf plot showing the number of brine shrimp hatched in a 6% salt solution.

Answer the questions below. Then, check your answers.

FIGURE 13.2 ORDERED STEM-AND-LEAF PLOT FOR NUMBER OF BRINE SHRIMP HATCHED IN A 6% SALT SOLUTION

```
Stem  Leaf
  0 | 7 7 8 8 9 9
  1 | 2 3 3 4 4 4 8
  2 | 2 2 4 6 8 8 9
  3 | 2 2 4 4 5 5
```

Key | 2 | 8 = 28

1. What values are represented by |0|9, |2|8, |3|4?

2. What is the minimum?

3. What is the maximum?

4. What is the range?

5. What is the mode?

6. What is the median of brine shrimp hatched?

7. Do the leaves make a pattern? If so, describe it.

> **Answers to Questions about Figure 13.2**
>
> 1. 9, 28, 34
> 2. 7
> 3. 35
> 4. 35 – 7 = 28
> 5. 14
> 6. 20
> 7. Leaves make a pattern that is similar to a rectangle. The same number of data points or leaves tends to be found within each value of the stem.

In an experiment, scientists are interested in comparing various groups. For example, brine shrimp were hatched in a 3% (Fig. 13.1) and a 6% (Fig.13.2) salt solution. How did the hatch rate compare? Which salt solution appeared to be best? One way that you can determine the best solution is by comparing the answers to the questions.

	3% Salt Solution	6% Salt Solution
What is the minimum?	7	7
What is the maximum?	57	35
What is the range?	50	28
What is the median?	33.5	20
What is the mode?	36	14
What is the shape?	Mound	Rectangular

For example, for the 3% salt solution both the median (33.5) and the mode (36) are higher than the median (20) and the mode (14) for the 6% salt solution. In addition, the hatch rate with the 3% salt solution covered a greater range.

FIGURE 13.3 BACK-TO-BACK STEM-AND-LEAF PLOTS FOR BRINE SHRIMP HATCHED IN A 3% AND 6% SALT SOLUTION

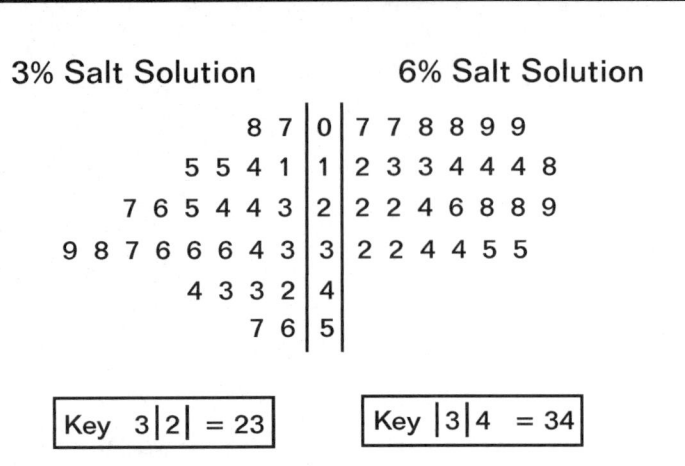

```
        3% Salt Solution              6% Salt Solution
                        8 7 | 0 | 7 7 8 8 9 9
                    5 5 4 1 | 1 | 2 3 3 4 4 4 8
                  7 6 5 4 4 3 | 2 | 2 2 4 6 8 8 9
              9 8 7 6 6 6 4 3 | 3 | 2 2 4 4 5 5
                      4 3 3 2 | 4 |
                          7 6 | 5 |

        Key  3|2| = 23                Key |3|4 = 34
```

A more effective way to compare the data is to make a back-to-back stem-and-leaf plot in which both sets of data are displayed together as shown in Figure 13.3. The stem is placed in the center and the leaves are placed on each side. Note that the smallest number or leaf is always placed closest to the stem. For the 6% salt solution, this appears normal for you are reading from left to right. However, for the 3% salt solution, you are reading in the opposite direction—from right to left. With the back-to-back plot, it is easier to see that the hatch rate is more spread out or dispersed with the 3% salt solution. Also, differences in the shapes of the distributions are more obvious.

Stem-and-Leaf plots have many benefits:

◆ Communicate the minimum, maximum, and range;

◆ Communicate the most typical values (median and mode);

◆ Provide a graphic display of every data point;

◆ Pictorially represent the spread or dispersion in the data;

◆ Enable one to determine if the data fit into one of the four common distribution patterns—Bell-shaped, U-shaped, J-shaped, or Rectangular shaped (see Figure 13.4)

Constructing a Stem-and-Leaf Plot

Now that you have analyzed several stem-and-leaf plots, it is time to make your own. Begin by looking at the data from the investigation on scooping and grabbing, which is shown in Table 13.1. By making a back-to-back stem-and-leaf plot, you can analyze the data effectively.

FIGURE 13.4 COMMON DATA DISTRIBUTION PATTERNS

a. Bell-shaped (Normal) Distribution: highs balance the lows.

```
1 | 0
2 | 3 3 8
3 | 2 5 6 7
4 | 3 3 4 5 5 8 8 9
5 | 2 5 5 8 9
6 | 2 4 7
7 | 2 3
8 | 1
9 |
```

b. U-shaped (Bi-Model) Distribution: two groups of relative equal frequency; if studied separately, each group may be bell-shaped.

```
1 | 0 2 2 7 8 9
2 | 3 4 5 6 6
3 | 0 4 5
4 | 2 4
5 | 1 6 7
6 | 2 3 4 5 8
7 | 1 1 2 5 7 8 8
8 |
9 |
```

c. J-shaped Distribution: shows that there is probably a limit to the values of the data.

```
1 | 0 2 2 4 5 7 7 8 9
2 | 1 3 3 6 7 7
3 | 0 4 5 6 8
4 | 3 3 4 5
5 | 2 3 4
6 | 7 8
7 | 2
8 | 2
9 | 0
```

d. Rectangular-shaped distribution: uniform distribution of data with values evenly distributed over a range.

```
1 |
2 |
3 | 0 3 3 4 8
4 | 2 3 4 4 7 8
5 | 1 2 4 8 8
6 | 3 4 5 5 9 9
7 | 2 3 4 7 7
8 |
9 |
```

Step 1. Determine the minimum and maximum values for the data. For grabbing, the values are 11 and 53. For scooping, the values are 22 and 77. To determine the values for the stem, use the minimum and maximum from both sets of data.

| Minimum | 11 | Stem = 1 |
| Maximum | 77 | Stem = 7 |

Making Stem-and-Leaf Plots

Step 2. In the center of the page, write the stem vertically. Draw dark lines on each side of the stem.

```
 |1|
 |2|
 |3|
 |4|
 |5|
 |6|
 |7|
```

Step 3. Make an unordered stem-and-leaf plot. Unordered means that the values do not occur in numerical order, such as 0 to 9. You can just enter each data point as you read it from the data table. On the right side of the stem, write the unordered "ones place" values of the data you collected for scooping.

Scooping

```
1|
2| 6 2
3| 0 3 6 8 3 7
4| 3 1 9 1 6 1 7
5| 6 5 3 0 3 7 7 7 8
6| 3 9 7
7| 7 0 0 4
```

Key | 4 | 6 = 46

Now, repeat the procedure on the left side using the data for grabbing.

Grabbing   Scooping

```
              1|
  6 3 2 5 8 7 1|1
      1 2 0 4 5 3 9 3|2| 6 2
    9 3 8 3 7 5 3 6 4 3|3| 0 3 6 8 3 7
              8 7 3|4| 3 1 9 1 6 1 7
                3 2 2|5| 6 5 3 0 3 7 7 7 8
                     |6| 3 9 7
                     |7| 7 0 0 4
```

Key 4|3 = 34   Key | 4 | 6 = 46

154   Chapter 13

Step 4. Make an ordered stem-and-leaf plot by arranging the leaves in order from the "inside to the outside." This time you will need to arrange them from 0 to 9. Begin with the data on the right side of the stem, the scoop data. Place the smallest number next to the stem and move outward. Order the numbers from left to right

Scooping

1	
2	2 6
3	0 3 3 6 7 8
4	1 1 1 3 6 7 9
5	0 3 3 5 6 7 7 7 8
6	3 7 9
7	0 0 4 7

Key |4|6 = 46

Next, repeat the process to order the grabbing data on the left side of the graph. Remember, the smallest numbers are closest to the stem! You will be ordering the numbers from right to left. This is like writing the numbers backward so be careful.

Key 4|3| = 34

Grabbing   Scooping

	8 7 6 5 3 2 1	1	
	9 5 4 3 3 2 1 0	2	2 6
9 8 7 7 5 4 3 3 3		3	0 3 3 6 7 8
	8 7 6 3	4	1 1 1 3 6 7 9
	3 2 2	5	0 3 3 5 6 7 7 7 8
		6	3 7 9
		7	0 0 4 7

Key |4|6 = 46

Be sure to include a key to communicate the values of the stem and leaves. For example, on the right or scoop side of the display, the key is |4|6 = 46. On the left or grab side of the display, the key is 4|3| = 34. Remember, the leaf and stem will be in opposite order for the two sides of the stem-and-leaf display.

Making Stem-and-Leaf Plots 155

Writing about Stem-and-Leaf Plots

Earlier you answered questions such as the following about stem-and-leaf plots:

◆ What are the minimum, maximum and range?

◆ What is the median?

◆ What is the mode?

◆ What is the shape of the data?

Now, you need to put your answers into sentences. After analyzing each set of data individually, you also need to make some comparisons. For example, how are the grabbing and scooping data alike? Different? When you began the experiment, you hypothesized which technique would be better. Did the data support your hypothesis?

Below is a paragraph we wrote about *Investigation 13.1, Scooping or Grabbing?* We had hypothesized that grabbing was the best technique.

> From 11 to 53 pieces of shell pasta were grabbed with the range being 42 pasta pieces. In contrast, 22 to 77 shell pasta pieces were captured when a scooping technique was used and the range was 55 shell pasta pieces. The mode for scooping (57) was greater than the mode for grabbing (33). Also, the median for scooping (50) was greater than the median for grabbing (33). Both sets of data had a general bell shape; however, in both cases the "highs" did not exactly balance the "lows." The data did not support the hypothesis that grabbing would be the best technique.

Multiple Data Sets

In many experiments, you compare more than two treatment groups. Can you still use stem-and-leaf plots? Yes, but you will need to make a stem-and-leaf plot for each set of data rather than using a back-to-back plot. In making the stem-and-leaf plots be sure to use the same values in the stem and to record the leaves on the same side. For example, if you tested the effectiveness of 0%, 3% and 6% salt solution in hatching brine shrimp you could display the data as shown in Figure 13.5. To write a paragraph, summarize your findings for each set of data, compare the distributions, and state if the data supported your hypothesis.

FIGURE 13.5 IMPACT OF PERCENTAGE OF SALT SOLUTION ON NUMBER OF BRINE SHRIMP HATCHED

0% Salt Solution

```
0 | 0 0 0 0 1 1 2 3 3 3 3 4 6 7 8 9
1 | 0 0 1 1 2 4 4 4 4 6
2 |
3 |
4 |
5 |
```

Key | 3 | 6 = 36

3% Salt Solution

```
0 | 7 8
1 | 1 4 5 5
2 | 3 4 4 5 6 7
3 | 3 4 6 6 6 7 8 9
4 | 2 3 3 4
5 | 6 7
```

6% Salt Solution

```
0 | 7 7 8 8 9 9
1 | 2 3 3 4 4 4 8
2 | 2 2 4 6 8 8 9
3 | 2 2 4 4 5 5
4 |
5 |
```

In a 0% salt solution, the number of hatched brine shrimp varied from 0 to 16 with most of the values falling between 0 and 10. In contrast, a 3% salt solution appeared most effective with the number of shrimp varying from 7 to 57 and most of the values falling in the thirties. With a 6% salt solution, the number of hatched shrimp declined with values ranging between 7 and 35 and almost equal numbers falling in each interval.

The smallest median hatch rate (6.5 shrimp) occurred with a 0% solution. Maximum hatching occurred with a 3% solution where the median was 33.5 shrimp. At the higher salt concentration of 6%, the median hatch decreased to 20 shrimp.

For a 3% salt solution, the distribution was bell shaped; whereas, for a 6% salt solution it was more rectangular shaped. Based upon the range and the shape of the data, the 3% salt solution appeared most effective. This supported the hypothesis that there would be an optimum salt concentration for brine shrimp hatching.

Making Stem-and-Leaf Plots

Applying Your Skills

Now that you have learned about stem-and-leaf plots, conduct *Investigation 13.2, Powdered Drink Currents*, display the data and write a paragraph. You can also use the practice problem from an experiment that tested the air-tightness of name brand and bargain brand plastic wrap to demonstrate your skills. If you have a favorite experiment from Chapters 1 to 12, use stem-and-leaf plots to display the data.

Investigation 13.2

Powdered Drink Currents

What You Need

- 1 Liter or larger clear container
- Metric ruler
- Room temperature water
- Ice water
- Paper towel squares
- Powdered drink mix (instant tea or a dark colored flavor of a fruit drink such as strawberry or grape)
- Measuring spoons, Metric
- Clock or watch with second hand

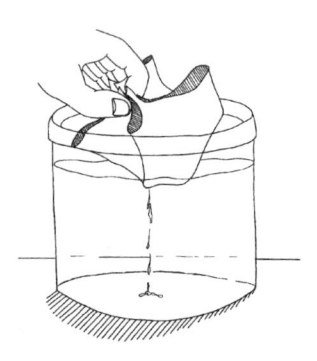

See Chapter 8, Experimenting Safely.

What You Do

1. Measure from the bottom of the container to the top. Place a mark approximately 8 cm below the top.
2. Fill the container with cold water to the mark.
3. Put 1mL of powdered drink mix in the center of the paper towel square. Fold the towel to make a "hobo bag." 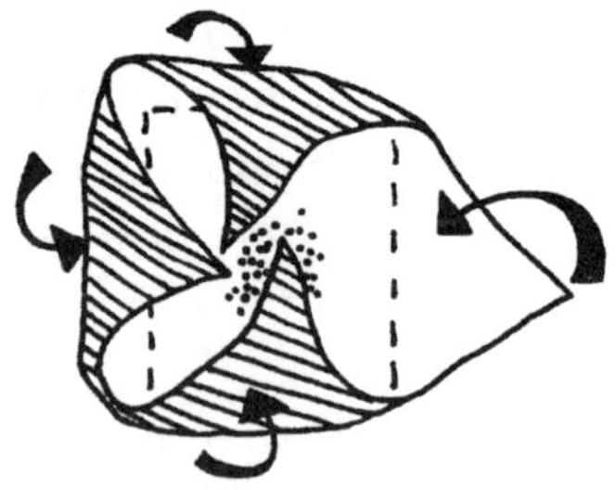
4. Place the tip of the "hobo bag" in the water. Watch the bottom of the bag carefully to see when a colored "stream" emerges. Time, in seconds, how long it takes for the colored "stream" to reach the bottom of the container. Make a data table, such as Table 13.1, to record your data.
5. Record your data in a class data table. Repeat Steps 1 to 4 as needed until at least 25 trials are recorded in the class data table.
6. Empty the cold water from the container. Fill to the mark with room temperature water. Repeat Steps 3-5.

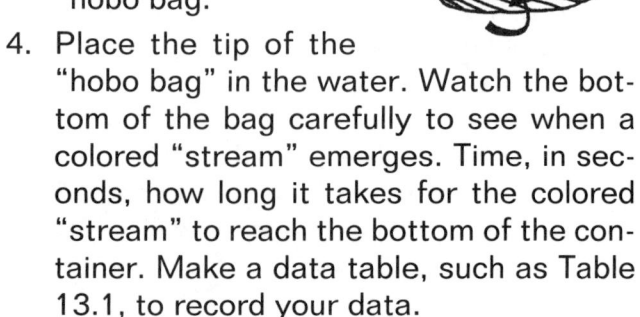

Searching the Web

For more information on Making Stem-and-Leaf Plots, use these search words or phrases:

"stem and leaf plots"
"stem-and-leaf"

to find websites such as this:
regentsprep.org/Regents/math/data/stemleaf.htm

Practice Problems

Making Stem-and-Leaf Plots

1. Plastic wrap is used to prevent food from drying out. Plastic wrap must be air tight to be effective. In order for seeds to sprout, oxygen is required. If seeds are placed in airtight plastic wrap, they will not sprout.

   Students decided to investigate the effectiveness of a name brand and a bargain brand of plastic wrap by comparing the germination rate of pea seeds. Fifty seeds were rolled in a damp paper towel. The paper towel roll was then wrapped in plastic wrap. After four days, the packets were unwrapped and the number of germinated seeds counted.

   Make a back-to-back stem-and-leaf plot of the data below. How did the seed germination rates compare? Which was the most effective plastic wrap? Why? Write a paragraph to summarize your findings.

Data for Practice Problem 1 The Effect of Type of Plastic Wrap on the Number of Germinated Seeds			
\multicolumn{4}{c}{Type of Plastic Wrap}
\multicolumn{2}{c}{Name Brand}	\multicolumn{2}{c}{Bargain Brand}
6	8	24	28
14	17	42	34
21	20	32	37
14	16	37	14
7	5	35	43
14	9	36	33
16	23	34	45
24	20	23	36
12	17	19	15
9	9	34	27
11	23	31	36
24	20	41	38
21	14	34	24
8	15	18	30
8	7	26	43

160   Chapter 13

Making Boxplots or Box and Whisker Diagrams

Graphic Displays, Quartiles, and Data Spread

Lava lamps, twirling liquid batons, and vinaigrette salad dressings—what do they have in common? All contain mixtures of substances such as oil, water, vinegar, alcohol, pigments, and solid particles. The inability of oil and water to mix is well known. Substances, such as soap and vinegar, are used to promote mixing. But, what about the movement of water drops through oil? Would the drops be affected by the presence of detergent? Carol and Sam decided to find out. They used the following materials:

- 100 mL graduated cylinder
- Ruler
- Water at room temperature
- Liquid dishwashing detergent
- Eyedropper
- Clock with second hand
- Liquid cooking oil
- Marker or tape
- 100 mL beaker

Sam and Carol poured 100 mL of cooking oil into a 100 mL graduated cylinder. Then they filled an eyedropper with water. Holding the dropper just above the surface of the oil, they released 1 drop of water into the oil. They measured the number of seconds that it took for each of 26 drops of water to fall to the bottom of the graduated cylinder filled with water. They then repeated the process using water that had a small amount of dishwater detergent dissolved in it (1.5 mL detergent in 100 mL water).

The data Sam and Carol collected for the descent of water and soapy water drops through oils is shown in Table 14.1, *The Effect of Detergent Dissolved in Water on the Time for Water Drops to Descend Through Cooking Oil.*

TABLE 14.1 Time for Water Drops with and without Detergent to Descend through Cooking Oil

Type of Water	
Without Detergent	With Detergent
7	5
7	5
8	6
9	6
9	6
12	7
13	7
13	7
15	8
15	8
17	8
17	8
18	9
18	9
18	9
18	9
18	9
19	10
20	11
20	11
21	11
21	11
22	12
22	12
23	13
24	13

It is difficult to see patterns in data when you have lots of values, even when they are arranged in order, as in Table 14.1. As you learned in Chapter 13, back-to-back stem-and-leaf plots are helpful in seeing patterns in data and making comparison between experimental groups. In this chapter you will learn how to display data using a type of graph called boxplots or box and whisker diagrams. To make a boxplot you will apply your knowledge of medians, which you learned in

Chapter 9. With boxplots, you will be able to describe additional aspects of the data including values below which 25%, 50% and 75% of the data fall.

> **Remember**
>
> To find the **median**, put the data in order and find the value in the middle.

From Stem-and-Leaf Plots to Boxplots

Some substances, such as rubber, can change their shape when a force is applied. Similarly, when the force is removed they return to their original shape. Substances with this property are said to be resilient. Both natural and synthetic rubber are resilient and are used to manufacture items such as balls, tires, and seals. Temperature affects the resiliency of objects and its impact is important when wide temperature changes are expected as in space and undersea exploration.

After reading that a cold rubber O ring's lack of resiliency was the cause of the Challenger Spaceship's explosion, Tim became interested in the time required for a rubber object to return to its original shape after exposure to different temperatures. He attached a C-clamp as tightly as possible on a handball to deform it and then placed the clamped handball in a container of water at a specific temperature for 10 minutes. After removing the handball from the water, he released the C-clamp and measured the time required for the ball to return to its original shape. The data Tim collected on the time for handballs to return to their original shape after immersion in 0°C and 15°C water are displayed as back-to-back ordered stem-and-leaf plots in Figure 14.1.

From the stem-and-leaf plots, you can see that the handballs exposed to 15°C returned more quickly to the original shape than the handballs exposed to 0°C. If you were observant, you

FIGURE 14.1 BACK-TO-BACK ORDERED STEM-AND-LEAF PLOTS FOR THE EFFECT OF TEMPERATURE ON THE TIME FOR HANDBALLS TO RETURN TO ORIGINAL SHAPE

0 degrees Celsius		15 degrees Celsius
	1	1 2 3 4 4 4 4 5 5 6 6 6 6 8 9
	2	0 1 1 1 3 4 4 4 5 6
9 9 6 6 5 3 3 0 0	3	
8 6 5 0 0 0	4	
6 5 5 5 3 2 2 1 0 0	5	

Key 6|4| = 46   Key |2|0 = 20

Making Boxplots or Box and Whisker Diagrams

also noticed some features of the stem-and-leaf plot that were not present before. For example, some of the leaf values are shaded and some have horizontal lines beneath them. These symbols are related to the median.

Look at the 25 values that Tim collected for 15° C. Find the median or middle value.

If you said 16 you are correct, for there are 12 values below and 12 above. Because you had an odd number of data points, the shading of 16 shows that it is exactly in the middle. Now, find the median for the half of the data below 16. Because there is an even number of values, the median (4) is between the sixth and seventh number and is shown as a line beneath the two 4's. Similarly, the median for the upper half of the data is 22 and is shown as a line beneath the values for 21 and 23. For practice, look at the data for 0° Celsius and explain what the lines and shaded value represent.

A boxplot is a graphic method of displaying data that is based upon five important points in the data.

Lower extreme	Minimum value
Lower quartile (Q_1)	Number below which 25% of the values fall
Median (Q_2)	Number that divides the data into half, with 50% of the values falling above the number and 50% below
Upper quartile (Q_3)	Number below which 75% of the numbers fall
Upper extreme	Maximum value

In addition the term 'interquartile' is used to refer to the middle 50% of the values and is mathematically expressed as $Q_3 - Q_1$. As shown in the diagram below, each of the points in a data set is displayed in a specific part of the boxplot.

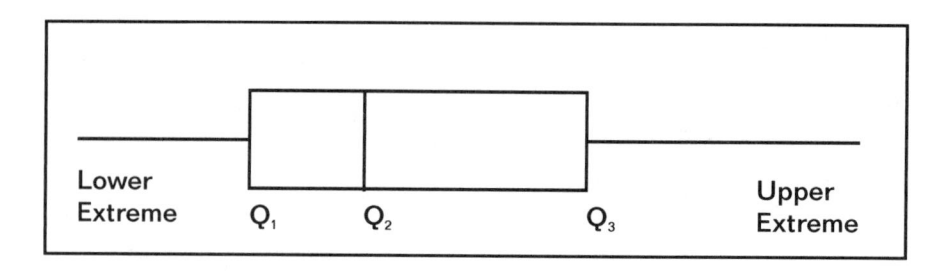

164 Chapter 14

Each whisker and each part of the box represents 25% of the data. If you were using a box plot to display 40 pieces of data, each whisker and each part of the box would represent 10 pieces of data, regardless of the length of the whisker or part of the box. The length of the whiskers and the parts of the box tell you how similar or different the data are. Ten pieces of data represented by a short whisker are closer in value to each other than are 10 pieces of data represented by a long whisker. Data represented by short whiskers and boxes are not as spread out as data represented by long whiskers and boxes. Long or short, each whisker and each part of the box represents the same amount of the data.

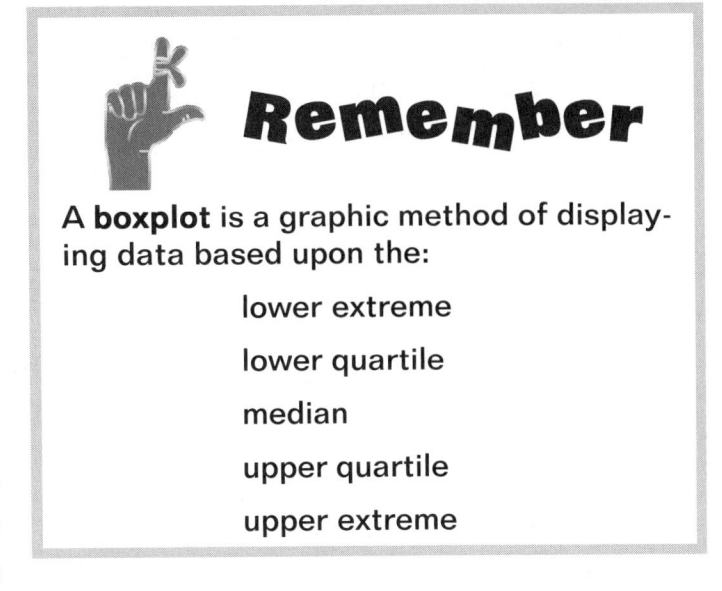

> **Remember**
>
> A **boxplot** is a graphic method of displaying data based upon the:
>
> lower extreme
>
> lower quartile
>
> median
>
> upper quartile
>
> upper extreme

Interpreting Boxplots or Box and Whisker Diagrams

By looking at the definitions and diagrams, you can interpret boxplots of data. To practice, let us look at Figure 14.2 that shows boxplots for the effect of temperature on the time for handballs to return to their original shape.

FIGURE 14.2 BOXPLOTS FOR EFFECT OF TEMPERATURE ON THE TIME FOR HANDBALLS TO RETURN TO ORIGINAL SHAPE

Time (sec) to Return to Original Shape

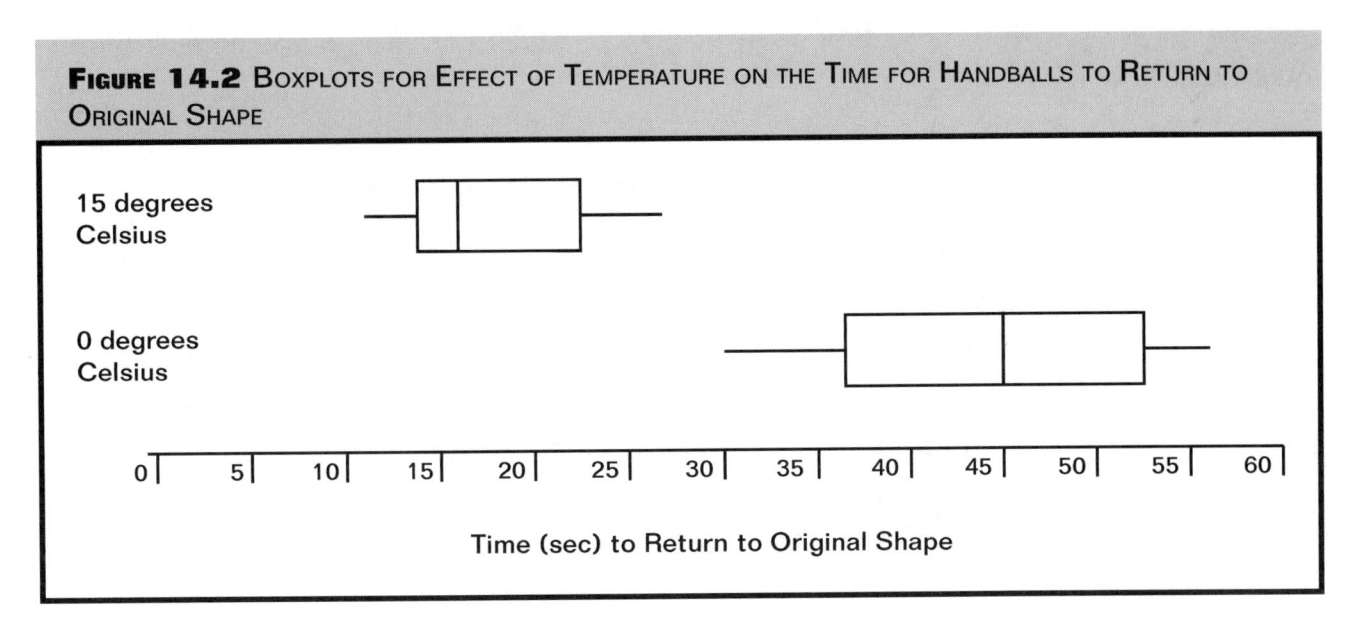

Let us begin by looking at the boxplot for handballs exposed to 0°C, which is shown on the bottom right of the display. By looking at the scale on the horizontal axis, answer the following questions.

1. What is the lower extreme? The upper extreme?

2. What are Q_1, Q_2 (median), and Q_3?

3. Below which value do 25% of the handballs fall? 50% of the handballs? 75% of the handballs?

4. What is the range of the interquartile (the middle 50%) of the handballs? Are these middle values for handball resiliency symmetrically distributed (evenly spread out)? How can you tell? (Hint: Look at the box.)

5. Where do you have the greatest dispersion or spread in the data: the lower quartile or the upper quartile? How can you tell? (Hint: Look at the length of the lines or whiskers.)

6. Because you were estimating from a graph, your numbers may not be exact. How could you use the stem-and-leaf plot to obtain more specific information?

After you have answered the questions, check your answers. If you need additional practice, then answer the same questions for the boxplot of handballs exposed to 15°C, which is located in the upper left of the display. Check your answers on page 167.

Constructing Boxplots or Box and Whisker Diagrams

Now that you know how to find medians and to interpret boxplots, it is time to construct boxplots for a set of data. As an example, we will use the data Sam and Carol collected for their experiment, *The Effect of Detergent Dissolved in Water on the Time for Water Drops to Descend through Cooking Oil*. The back-to-back stem-and-leaf plots for the data are displayed in Figure 14.3; look at the stem-and-leaf plot as you follow the steps below. You will easily see how we obtained the values to use in constructing the boxplots.

Answers to Questions about Figure 14.2 – Handballs Exposed to 0°C

1. Lower extreme = 30; Upper extreme = 56
2. $Q_1 = 36$; Q_2 or median = 45; $Q_3 = 52$
3. 25% of values fall below 36; 50% of values fall below 45; 75% of values fall below 52
4. The interquartile (the middle 50% of the values) is between 36 and 52; values are slightly more spread out between Q_1 and Q_2 than between Q_2 and Q_3 and therefore the lower half of the box is slightly longer.
5. The greatest dispersion (spread) is in the lower half because the line or whisker is longer.
6. From the stem-and-leaf plot you can determine the exact values; whereas, for the boxplot you must estimate from the scale of the graph. Look back at Figure 14.1 and see how well your estimated values compared with the actual.

Answers to Questions about Figure 14.2 – Handballs Exposed to 15°C

1. Lower extreme = 11; Upper extreme = 26
2. $Q_1 = 14$; Q_2 or median = 16; $Q_3 = 22$
3. 25% of values fall below 14; 50% of values fall below 16; 75% of values fall below 22
4. The interquartile (the middle 50% of the values) is between 14 and 22; values are very much more spread out between Q_2 and Q_3 than between Q_1 and Q_3 resulting in the upper half of the box being much longer
5. The dispersion (spread) in the upper quartile is slightly more because the line or whisker is slightly longer.
6. From the stem-and-leaf plot you can determine the exact values; whereas, for the boxplot you must estimate from the scale of the graph. Look back at Figure 14.1 and see how well your estimated values compared with the actual.

FIGURE 14.3 BACK-TO-BACK STEM-AND-LEAF PLOTS FOR THE EFFECT OF DETERGENT DISSOLVED IN WATER ON THE TIME FOR WATER DROPS TO DESCEND THROUGH COOKING OIL

```
        Water Drops without Detergent              Water Drops with Detergent

                         9 9 8 7 7 | 0 | 5 5 6 6 6 7 7 7 8 8 8 8 9 9 9 9 9
           9 8 8 8 8 8 7 7 5 5 3 3 2 | 1 | 0 1 1 1 1 2 2 3 3
                   4 3 2 2 1 1 0 0 | 2 |

                    Key  7|0| = 7              Key |1|1 = 11
```

Step 1. Find the median or mid-point of the data. Because there are 26 values, the median will be the point halfway between the 13th and 14th value. This value is underlined in the stem-and-leaf plot.

Median (Q_2) for water without detergent   18

Median (Q_2) for water with detergent   9

Step 2. Find the median of the lower half of the data (Q_1). Because there are 13 values, the median will be the 7th value; there will be 6 numbers below it and 6 numbers above it. This value is shaded in the stem-and-leaf plot.

Lower quartile (Q_1) for water without detergent   13

Lower quartile (Q_1) for water with detergent   7

Step 3. Find the median for the upper half of the data (Q_3). Again, because there are 13 values, the median will be the 7th value. This value is shaded in the stem-and-leaf plot.

Upper quartile (Q_3) for water without detergent   20

Upper quartile (Q_3) for water with detergent   11

Step 4. Find the extreme values of the data; that is the minimum and maximum values.

Extreme values for water without detergent   Minimum = 7
                                              Maximum = 24

Extreme values for water with detergent   Minimum = 5
                                              Maximum = 23

Step 5. Now that you have the values that you need to make the box and whiskers, you need to determine an appropriate interval scale for graphing the boxplots. To do this, look at the extreme values in Step 4 and determine the smallest number and largest number that you will need to graph. Use the techniques previously described in Chapters 10 and 11 to determine the scale for the horizontal number line or axis.

$$\frac{\text{Maximum} - \text{Minimum}}{5} = \frac{24 - 5}{5} = \frac{21}{5} = 4.2$$

Draw a horizontal number line beginning at 0 and ending at 30. Remember that you always go one interval below the lowest value and one interval above the largest value. Remember to label the horizontal number line – "Time to descend (sec)." For an example, see Figure 14.4, *Boxplots for The Effect of Detergent Dissolved in Water on the Time for Water Drops to Descend through Cooking Oil.*

Step 6. Write a label, such as "Water Drops without Detergent" to the left of the number line. Then, use the information in Steps 1 to 4 to make a boxplot. Begin by drawing a vertical line for the median (Q_2). Next, draw vertical lines for Q_1 and Q_3. Connect the vertical lines to form a box. Finally, draw a horizontal line or whisker from the lower end of the box to the minimum. Similarly, draw a whisker to the maximum. If you need help, look at Figure 14.4.

Step 7. Above the label and boxplot for "Water Drops without Detergent" make a label and boxplot for the other set of data, "Water Drops with Detergent." Follow the procedures in Step 6.

FIGURE 14.4 BOXPLOTS FOR: THE EFFECT OF DETERGENT DISSOLVED IN WATER ON THE TIME FOR WATER DROPS TO DESCEND THROUGH COOKING OIL

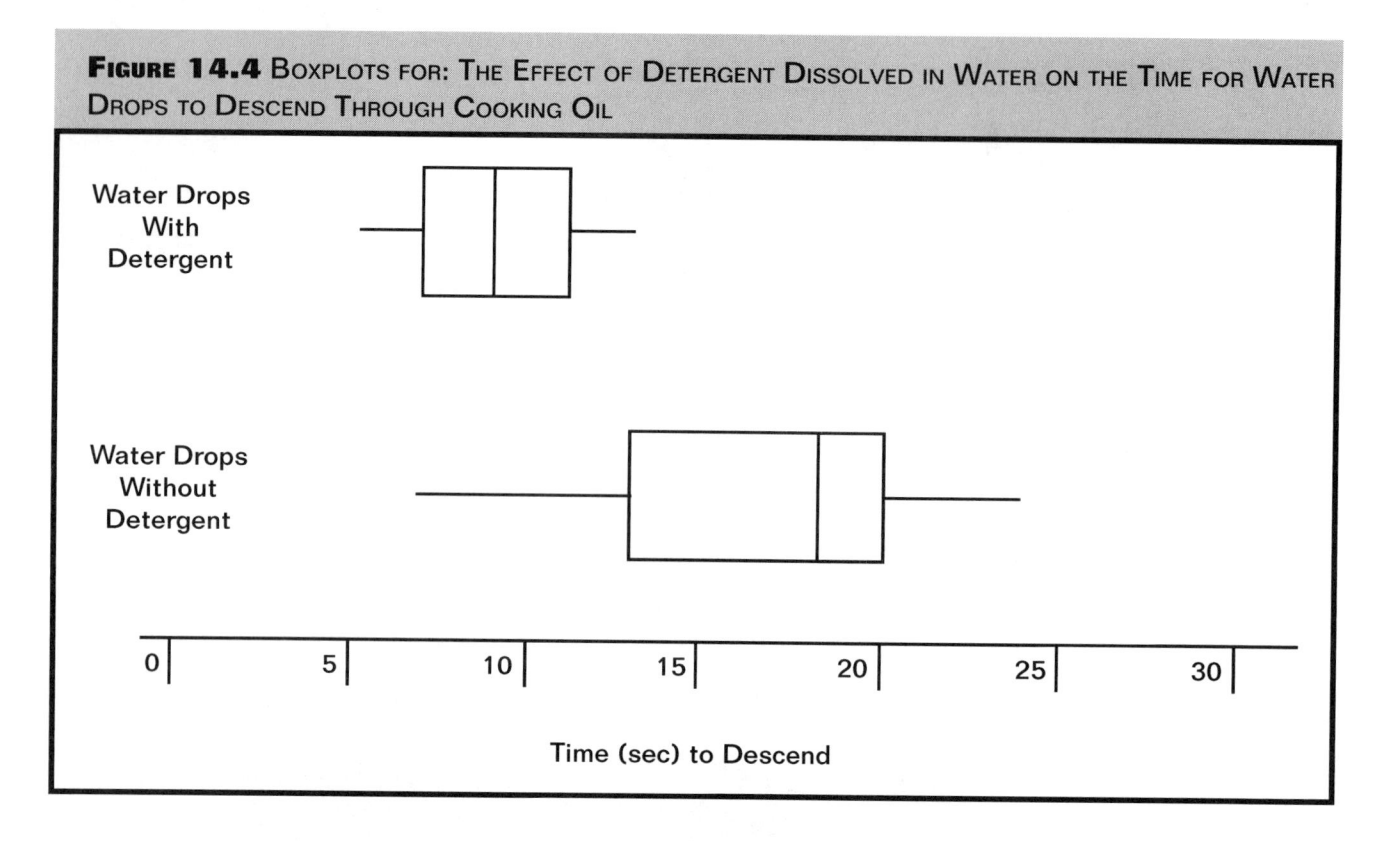

Step 8. Check the boxplots against the values you calculated in Steps 1 to 4 to be sure that you graphed correctly. Also, double check that you labeled the box plots and the horizontal axis or number line correctly. Everything OK? Congratulate yourself!

Writing about Boxplots

Just as with data tables, graphs, and stem-and-leaf plots it is important to write a paragraph summarizing the information displayed in boxplots. Important information for the paragraph includes:

◆ What are the medians? How do they compare?

◆ What are the interquartile ranges, or difference between the upper quartile (Q_3) and the lower quartile (Q_1)? How do they compare?

◆ Are the values symmetrical (evenly spread) in the box or middle 50%? How do the interquartile boxes compare?

◆ Are the values symmetrical (evenly spread) in the upper and lower extremes? How do the whiskers compare?

◆ Do the data support the hypothesis?

170 Chapter 14

Below is a paragraph we wrote about *Investigation 14.1, Oily Descents*.

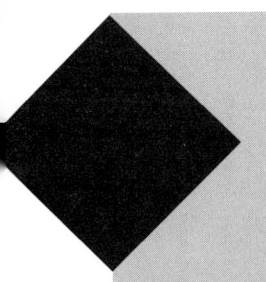

> The time for water drops, with and without detergent, to fall through oil was investigated. The data were displayed in boxplots (see Figure 14.4). The median fall time was half as long for water with detergent (9 sec) as in plain water (18 sec). With detergent, the data were also more spread out and the interquartile range (7) was almost twice as much as with plain water (4). With detergent, the descent values were spread out evenly throughout. In contrast, the fall times in water without detergent were not evenly spread out. A much greater spread was shown in the lower 50%. Both the lower half of the box and the whisker for the lower quartile were longer. The data supported the hypothesis that because detergent helps oil and water mix, it would also help the water move through the oil.

Displaying Multiple Boxplots

Although the examples in this chapter have involved a comparison between only two boxplots, you can make boxplots for as many experimental groups as you have. Just place appropriate labels on the left side of the graphic display and stack the boxplots, one above another.

As an example, consider an experiment to investigate the impact of different amounts of sugar on the strength of a glue-like paste. Sugar, rice, starch, water, and gum Arabic were combined to make the paste. The recipe was modified to use the normal amount of sugar, one-half as much sugar, and twice as much sugar. The various pastes were used to glue pine boards and after a drying period the force in dynes (d) needed to break the boards apart was determined.

Because the independent variable (the amount of sugar in the paste) contained three levels it is appropriate to make a graphic display containing three boxplots. In Figure 14.5, the boxplots are displayed. As with any other graphic display, you also need to write a paragraph and make comparisons among the various boxplots. Use the questions previously provided as a guide.

> The effect of different amounts of sugar on the strength of paste is shown in a set of boxplots (see Figure 14.5). As the amount of sugar increased, the median force needed to break the glue bond increased, for example 2 g sugar, 4 g sugar (control) and 8 g sugar. The dispersion or spread increased as the amount of sugar increased. For all three amounts of sugar, the data were

FIGURE 14.5 BOXPLOTS FOR: THE EFFECT OF VARIOUS AMOUNTS OF SUGAR ON THE STRENGTH OF PASTE AS MEASURED BY FORCE (DYNES) REQUIRED TO SEPARATE

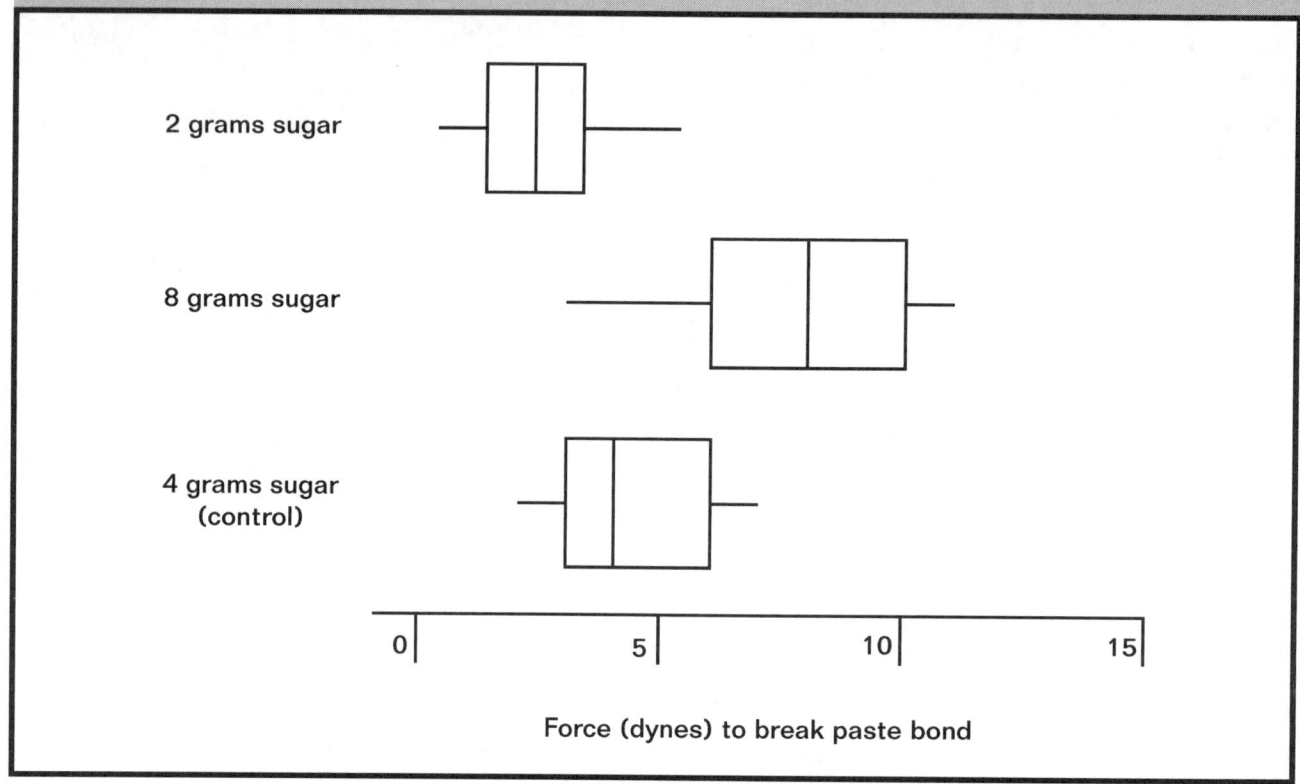

Force (dynes) to break paste bond

fairly symmetrical (evenly spread out). Exceptions were that the control group had a wider range in the upper 50%, the 8 g of sugar had a wider range in the lower quartile, and the 2 g of sugar had a slightly greater range in the upper quartile. The data did not support the hypothesis that the control would be the strongest and that changing the recipe would result in weaker pastes.

Applying Your Skills

To practice your new skills in making boxplots, conduct **Investigation 14.1, Quilted Strength**, and **14.2, Straw Javelins**. Then for each experiment display the data you collected and write a paragraph. You can also use the data from **Investigation 9.2 Paper Worms** or the data from Practice Problem 1 at the end of this chapter to demonstrate your understanding of stem-and-leaf and box plots. If you have another favorite experiment from Chapters 1 to 13, use box plots to display the data.

Investigation 14.1

Quilted Strength

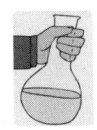

What You Need

- 250 mL beaker or a plastic cup (8 oz)
- Quilted paper towel squares
- Non-quilted paper towel squares
- Rubber bands
- Water
- Pennies (or washers)
- Pie plate or pan
- Graduated cylinder

See Chapter 8, Experimenting Safely.

What You Learned

1. Make boxplots to display the data for the wet quilted and non-quilted paper towels.
2. Write a paragraph summarizing your findings.

What You Do

1. Place a quilted paper towel square in a pie plate or pan. Pour 50 mL of water in the center of the quilted paper towel.
2. Place the wet quilted paper towel over the top of a 250 mL beaker or plastic cup. Put a rubber band around the top of the cup so that the paper towel is stretched tight and flat. 
3. Put pennies on the paper towel until it breaks. Count the number of pennies. Record the number of pennies in a class data table. Repeat Steps 1 to 3 until there are at least 25 trials shown in the class data table.
4. Repeat Steps 1 to 4 using a non-quilted paper towel.

Making Boxplots or Box and Whisker Diagrams

Investigation 14.2

Straw Javelins

WHAT YOU NEED

- Non-flexible plastic drinking straw
- Meter stick
- An area about 3 m wide by 9 m long in which it is safe to throw weighted and un-weighted plastic straws
- Masking tape
- Paper clip
- Safety goggles

See Chapter 8, Experimenting Safely.

WHAT YOU DO

1. Cut 7 pieces of masking tape 30 cm long. Place a strip of tape in the middle of the narrow end of the safe throwing area. Write **THROW LINE** on the strip of tape. Then place the other pieces of tape at 1 m intervals down the center of the area. Mark them 1 m, 2 m, etc.

2. Put on the safety goggles. Stand on the **THROW LINE**. Hold the plastic straw like a dart or a javelin in your dominant hand. Throw it as far and as straight as you can in the safe throwing area and measure the distance you threw the straw.

3. Enter your data in a class data table.

4. Repeat Steps 2 and 3 until there are at least 25 entries in the class data table.

5. Insert a regular paper clip into the front end of the straw to weight it. Attach the paper clip to the side of the straw. Repeat Steps 2-4 with the weighted straw.

WHAT YOU LEARNED

1. Make boxplots to display the data for the throws with your dominant hand and with your non-dominant hand.

2. Write a paragraph summarizing your findings.

174 Chapter 14

Searching the Web

For more information on Making Boxplots or Box and Whisker Diagrams, use these search words or phrases:

"boxplots"
"box and whisker"
"box and whisker diagrams"

to find websites such as this:
ellerbruch.nmu.edu/cs255/jnord/boxplot.html

Practice Problems

Making Boxplots or Box and Whisker Diagrams

1. In Lauren's math class the teacher turned the lights on and off frequently as he used the overhead projector. Lauren noticed that her classmates were constantly blinking their eyes, just as when they emerged from a dark movie theatre. Lauren had read that light affected the diameter of the pupil. She decided to investigate if the size of the pupil changed with the partial and full light in the classroom. Lauren hypothesized that the pupils would have a larger diameter in dim light.

   Below are the data Lauren collected after students had been in full classroom light for 15 minutes and in partial classroom light for 15 minutes.

   Order the data by making lists or back-to-back stem-and-leaf plots. Because the numbers are very small, the stem-and-leaf plot will be different from those in Chapter 13. The stem will be the "ones" place and the leaves will be the "tenths" place.

   Use boxplots to display the data. Write a paragraph to summarize your findings.

The Effect of Light on the Diameter (cm) of the Pupil	
Amount of Light	
Classroom Light	Partial Classroom Light
4.3	6.0
4.2	6.2
4.7	5.8
3.9	5.7
5.2	6.2
5.1	5.8
4.8	6.1
4.7	5.4
4.5	6.3
4.5	5.6
3.8	6.3
3.7	6.2
3.9	6.1
3.9	5.9
4.0	5.8
4.2	5.3
4.0	5.2
4.1	5.8
4.3	5.7
3.8	6.1
3.4	6.2
4.4	6.4
3.8	5.8
3.9	5.9
4.0	5.1

Presenting Your Experiment

Rules, Displays, Posters, and Presentations

Have you ever been stuck behind a big truck going up a steep hill? Steep hills and ramps are of special concern to truckers and highway engineers. When hills and ramps are studied in science classes they are called incline planes. Suppose you did a science project studying the effect of the steepness of an incline plane on the ability of a "truck" to climb it. Being a good truck driver requires lots of skills and being alert to avoid accidents. Stay alert! Avoid an accident. Stop the truck at the end of the ramp each time you run a trial in *Investigation 15.1, It's All Up Hill*.

Investigation 15.1

It's All Up Hill

What You Need

- Battery-operated 4WD toy truck
- Appropriate size batteries
- A flat shelving board at least 1.5 m long
- 5 × 5 cm square of cardboard
- 3 Number 2 fishing sinkers
- 1 piece of string 20 cm long

See Chapter 8, Experimenting Safely, Section C, Electricity.

What You Do

1. Construct a ramp so that it raises 25 cm in 1.5 m of ramp surface.
2. Place fresh batteries in the truck and tie the cardboard "sled" to drag behind the truck.
3. Place the truck at the bottom of the ramp and measure the time (sec) that it takes the truck to travel 1.5 m up the ramp's surface. Record the time (sec) in Table 15.1, and note the amount of slippage of the wheels in Table 15.2 on page 178.
4. Repeat the process with 1, 2, and 3 sinkers taped to the sled behind the truck.
5. Repeat Steps 3 and 4 four more times for a total of 5 trials.

TABLE 15.1 The Effect of Varying Cargo Mass on the Climbing Time of a Battery Powered Truck

Number of Sinkers	Time to Climb Ramp (sec) Trials					Mean Time to Climb Ramp (sec.)	Range (sec.)
	1	2	3	4	5		
0							
1							
2							
3							

TABLE 15.2 The Effect of Varying Cargo Mass on the Wheel Slippage of a Battery Powered Truck

Number of Sinkers	Slippage on a Ramp Trials					Mode	Frequency Distribution
	1	2	3	4	5		
0							
1							
2							
3							

Key: N = No slipping of wheels

M = Minor slipping of wheels (No pauses in forward motion)

S = Some slipping wheels (Slight pauses in forward movement)

L = Lots of slippage (Jerky forward motion)

Let's suppose you have written a report for the experiment "It's All Up Hill" as you learned to do in Chapter 12. Now you're ready to present your experiment in a competition such as a Science Fair or a State Junior Academy of Science, or are you?

Rules of Competition

Let's begin preparing for the competition by taking a short break. During the short break, how about a quick game of Squibbish? Before you can answer this question you need to know what Squibbish is...what special equipment is needed to play the game, what the rules are, and so on.

Similarly, you need to know the rules of the science competition you're entering before you can start preparing for your presentation. In fact, some competitions have rules you must follow even before you start your experiment.

For example, the International Science and Engineering Fair rules require you to have a set of forms filled out before you even purchase animals for use in a possible experiment. Every competition has its own rules that are published. Some competitions allow procedures and materials to be written as lists, while other competitions require procedures and materials be written in paragraph form. Many competitions require a written report describing your work. Competitions have definite limits on the length of the paper, specifying single or double spacing, and even the size of margins.

The point is simple. Get the rules for the competition you wish to enter, and, if possible, the instructions to the judges. It is important that you present your work in such a way as to emphasize the things that judges are looking for. In the Shot-Put event at a track meet, two things are important—that your feet stay in the circle and how far you can throw the shot. Your artistic form is not judged as it is in figure skating. Different competitions, different rules. KNOW THE RULES!

Now for the game of Squibbish...Just teasing. Let's get on with the task of presenting your project.

Science Fair Displays

The format of a science fair display board is similar to the components of a simple report: title, statement of problem, procedure (methods-materials), results, and conclusion. Figure 15.1 depicts the traditional position of these components on a science fair display board.

A model of the components of a 3-panel display for **Investigation 15.1, It's All Up Hill** follows. Use it as a model for your own display.

Title

A title may state the specific variables you investigated or may be worded creatively to capture the audience's attention.

- ◆ The Effect of a Truck's Sled Load on Climbing Time and Wheel Slippage
- ◆ The Effect of Sled Load on a Truck's Ability to Climb a Hill
- ◆ Tired Trucks: It's All Up Hill!

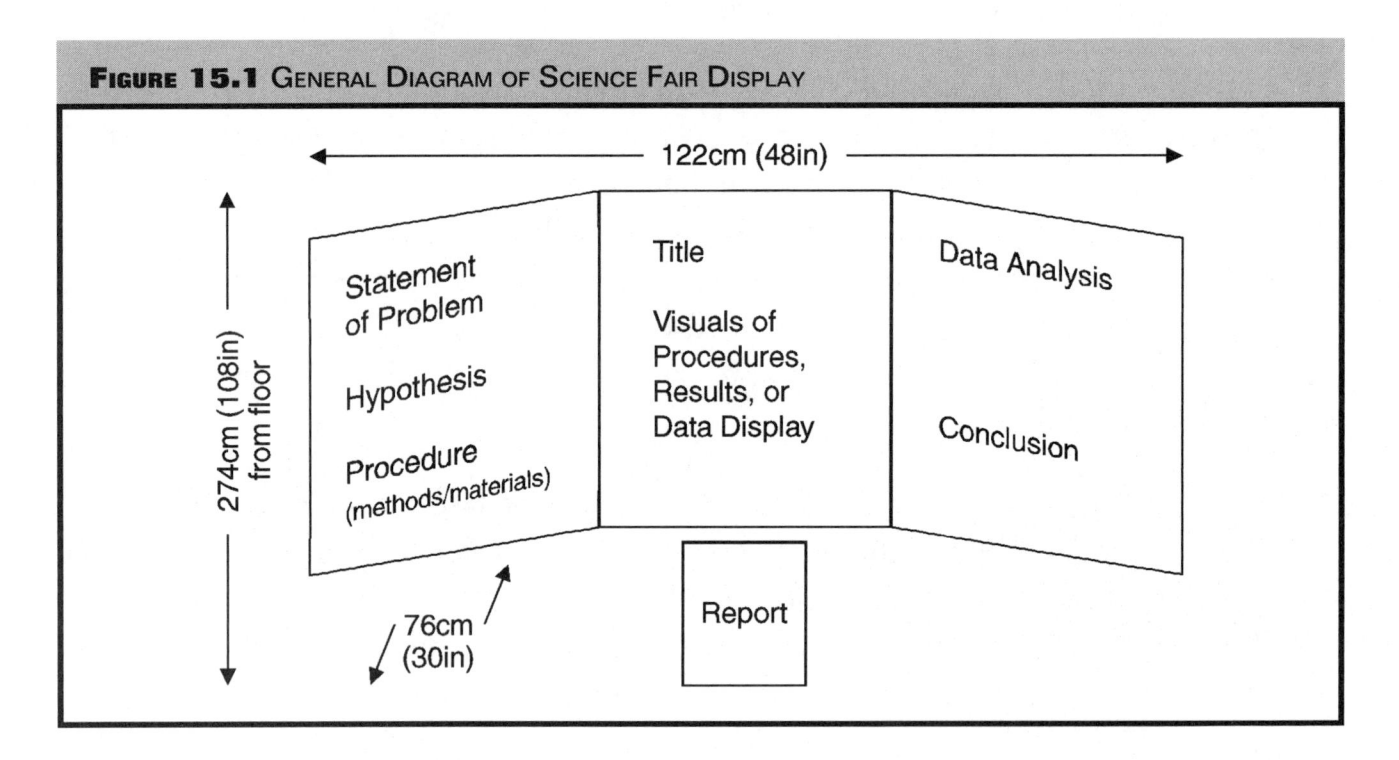

FIGURE 15.1 GENERAL DIAGRAM OF SCIENCE FAIR DISPLAY

[Diagram showing a tri-fold science fair display board with dimensions 122cm (48in) wide and 274cm (108in) from floor, with a depth of 76cm (30in). The left panel shows "Statement of Problem, Hypothesis, Procedure (methods/materials)". The center panel shows "Title, Visuals of Procedures, Results, or Data Display". The right panel shows "Data Analysis, Conclusion". A "Report" is placed in front.]

Statement of Problem

Tell your audience why you did the experiment, what you hoped to learn, and what you thought would happen. Be brief! Use essential information from the introduction section of your written report to write a paragraph, or simply list the three major things you want to communicate.

Reason
A truck dealer advertises that her trucks perform even better with heavy loads.

Purpose
To determine how sled load affects a truck's climbing time and wheel slippage.

Hypothesis

If the sled mass pulled by a battery operated truck is increased, then the time to climb an incline and the amount of wheel slippage will increase. If space allows, you can also include components of your experimental design diagram. Since you have already given the title and hypothesis, you don't have to repeat them.

IV: Mass of Load			
0 Sinker (control)	1 Sinker	2 Sinkers	3 Sinkers
5 Trials	5 Trials	5 Trials	5 Trials

DV: Time (sec) to climb ramp
               Wheel slippage (none, some, much, lots)
C: Truck—Jones Toy Company, Model 376A
               Batteries—Size "AA," 1.5 v., Ever-On
               Incline—Wooden board, 1.5 m, 25 cm incline
               Sinkers—Number 2, tear-drop shaped, lead
               Sled—5 cm x 5 cm, piece of Index card paper

Procedure

You can communicate your procedure as a list of steps or a paragraph. In either case, be brief! All you need to give are the most important features of your procedure. The details are in your written report.

> A truck, Jones Toy Company Model 376A, was battery powered by 1 "AA" size Ever-On 1.5 volt battery. The truck and attached cardboard 5 x 5 cm "sled" were placed on the bottom of a ramp with no weights placed on the "sled." The length of time (sec) it took the truck to pull the empty sled up a wooden ramp surface was measured. The ramp was inclined 25 cm per 1.5 m of ramp surface. A new battery of the same brand was put into the truck. The procedure was repeated with 1, 2, and 3 number 2 fishing sinkers taped to the sled. The whole experiment was repeated four more times for a total of 5 trials.

Presenting Your Experiment   **181**

Results

You will need to include data tables and graphs that state your findings. Typically, space will not allow you to include all your tables and graphs—you'll have to make decisions. Choose only the most important ones. If you're really limited for space, you could even eliminate data tables and include only graphs.

TABLE 15.3 Data Table — Time to Climb Ramp

Number of Sinkers	Time to Climb Ramp (sec) Trials					Mean Time to Climb a Ramp (sec.)	Range (sec.)
	1	2	3	4	5		
0	5	4	5	4	5	5	1
1	8	10	8	8	9	9	2
2	11	12	13	11	12	12	2
3	17	18	18	17	17	17	1

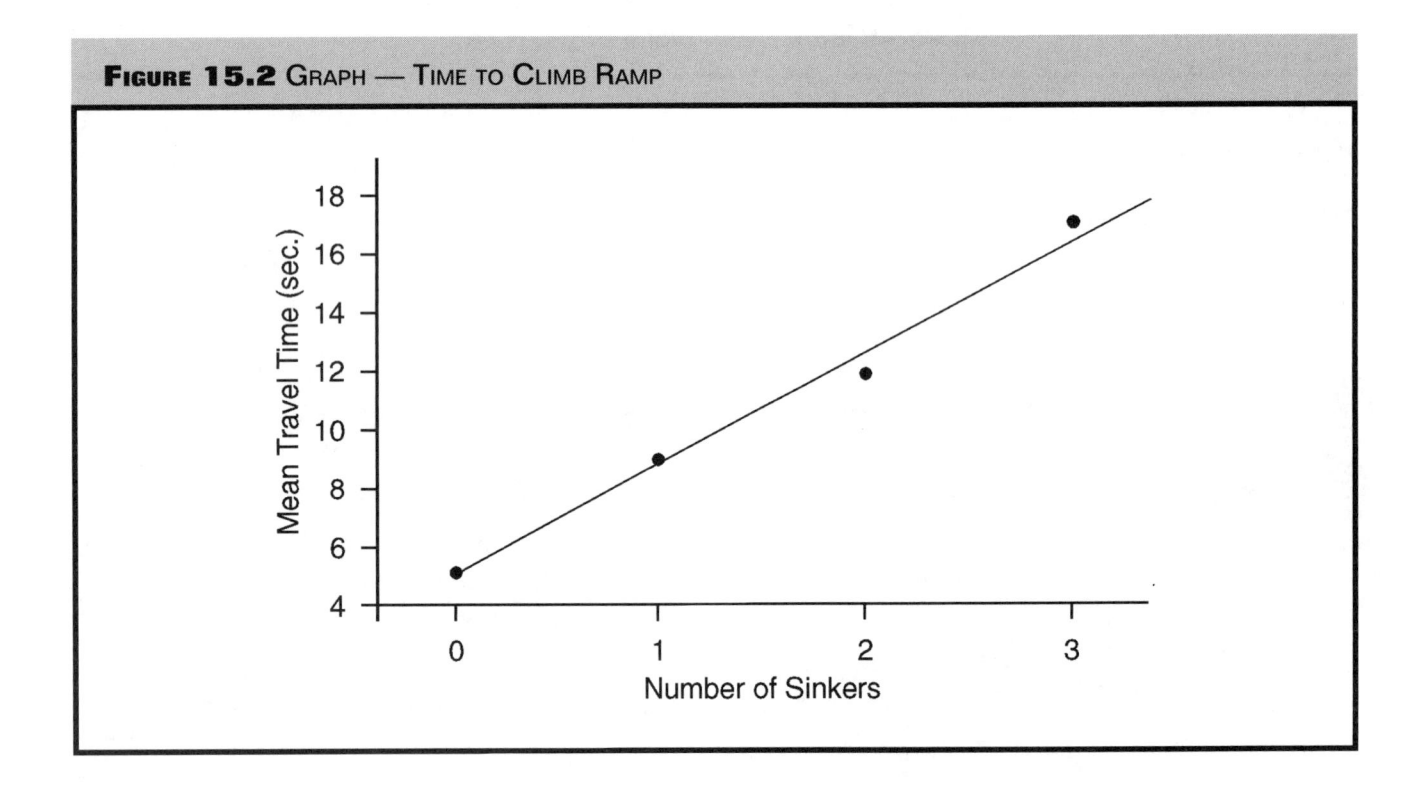

FIGURE 15.2 GRAPH — TIME TO CLIMB RAMP

Summary Sentence

As the number of sinkers on the sled increased, the time that it took the truck to pull the sinkers up the slope also increased.

TABLE 15.4 Data Table —The Effect of Mass on Sled on the Wheel Slippage of a Battery Operated Truck

Number of Sinkers	\multicolumn{5}{c}{Slippage in Climbing Ramp Trials}	Mode	Frequency Distribution
	1	2	3	4	5		
0	N	N	N	L	N	N	N: 4 M: 0 S: 0 L: 1
1	N	M	N	M	M	M	N: 2 M: 3 S: 0 L: 0
2	M	S	L	S	S	S	N: 0 M: 1 S: 3 L: 1
3	L	L	S	S	L	L	N: 0 M: 0 S: 2 L: 3

Key: N = No slipping of wheels
   M = Minor slipping of wheels (No pauses in forward motion)
   S = Some slipping wheels (Slight pauses in forward movement)
   L = Lots of slippage (Jerky forward motion)

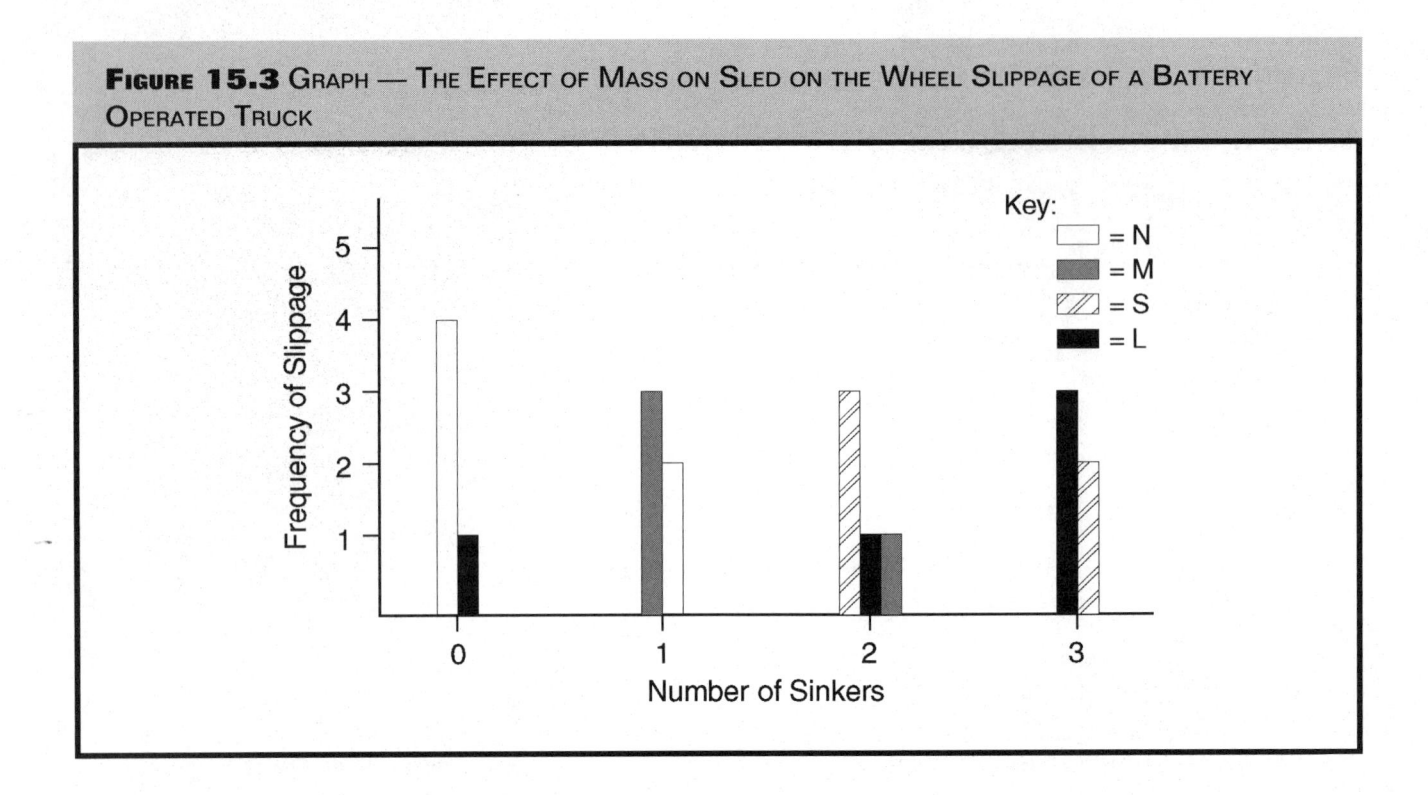

FIGURE 15.3 GRAPH — THE EFFECT OF MASS ON SLED ON THE WHEEL SLIPPAGE OF A BATTERY OPERATED TRUCK

Summary Sentence

As the number of sinkers on the sled increased, the amount of wheel slipping by the towing truck also increased.

Conclusion

Briefly state a summary of your findings, an explanation of results, and suggestions for improvements or other experiments. Display space will not allow a lengthy discussion as given in your written report.

Heavy loads on a truck sled increased the time required for the truck to climb the hill and the amount of wheel slippage. The data did support the hypothesis that truck performance would be reduced. Other experiments should be conducted with the load placed directly on the truck, using different types of road surfaces, and different steepnesses of the incline.

Tips for Constructing a Display

When you enter a science fair competition, you will need a display board. Most display boards are three panels "hinged" together so that they can stand up. Three ways to obtain a display board are:

1. Buy a three-panel display board from an office supply store, school supply store, or science fair materials supplier;
2. Construct one using:

   3 wooden panels (plywood, press board, paneling),
   4 hinges with screws,
   material to cover front surface of panels;

3. Construct one using 6 sheets of poster board (double wall construction is needed to give rigidity) and 1 roll of vinyl tape.

   If you make a trifold display (see Figure 15) from poster board, follow these steps:

   1. Cut a piece of tape exactly the length of the poster board and lay the tape sticky side up on a large flat surface.
   2. Place a piece of the poster board on the tape so that it covers ½ of the width of the tape.
   3. Place the second sheet of poster board on top of the first sheet so that all sides match.
   4. Fold the other half of the tape over the edge of the sheets of poster board.
   5. Similarly, tape the other three sides of the pair of poster boards.
   6. Repeat steps 1-5 for the four remaining pieces of poster board.
   7. Cut a strip of vinyl plastic tape double the length of the poster board. Lay it on a flat surface sticky side up.
   8. Place two of the panels side by side on the sticky surface.
   9. Lift the tape at the top and lay it over the seams. Smooth the tape carefully. Repeat to join the third panel to the other two. "Presto" you have a sturdy three panel display board.

Now it's time to put selected, important information on your display board.

◆ Neatness is important.

◆ The board must be read at a distance of one to two meters, so be sure the lettering is large enough.

Presenting Your Experiment 185

◆ Communicate your experiment clearly and briefly in the most attractive way possible. Plastic letter stencils of several sizes are available from art supply stores and the stencils can be reused for many projects.

◆ Plan your whole display ahead of time before you start writing on the surface. One mistake can ruin the appearance of the display. Many students type or print each component on a 5 x 8 in. card and glue the card neatly to the backboard. This way, if you make a mistake, you can throw the card away and your display board has only your best work on it! If you are going to use a computer to print items for use on your display board, use a large type size. Keep each printed message as brief as possible. Remember, in most cases you will be there to describe the details of your project to the judges.

Oral Presentations

You'll want your presentation to be as smooth and as convincing as the performances of your favorite actor or actress. Well, almost!

When you watch a TV show, the actors make the action look real. You get the impression that the actors are really angry, tense, or deeply in love. They seem to be saying the words as they think of them. The actors make you think the actions, words, and their emotions are real. Of course, the actors have rehearsed each scene for weeks or maybe months to get just the right effect.

Excellent presentations are carefully thought out. They are practiced many times and then enthusiastically presented. But, what should you say about your project? What do the judges need to be told?

Your oral presentation should state your:

1. Name
2. Grade and school (if this information is allowed by competition rules)
3. Project title
4. Reason (rationale) for doing the experiment
5. Experimental design diagram—information
6. Procedure

7. Results (data/graphs) and their meaning

8. Conclusions

9. Responses to judges' questions.

Looks like a lot? Look at your report. Most of the items listed above are in it. Now, get a pack of 5 x 8 cards. Write a large 1 in the upper right of a card and then in big letters print your name. On another card write a large 2 in the upper right corner and write, you guessed it, your grade and school's name. On the 3rd card in the upper left corner write ... You guessed this too. On cards 3-8 you should *outline* your talk. Don't plan on reading your cards to the judges. That would make it look like you don't know what you did. An outline of Step 5 may take 3 or more cards—5A, 5B and 5C ... If you drop the cards 30 seconds before the judges get to you, you must be able to put them back in order fast, hence the numbering.

Now, this next suggestion may sound weird, but do it. Stand in front of a mirror and read your cards in order from first to last. Now, read them and try to connect them as you go from card to card. Underline important words on each card. Keep practicing until your presentation is smooth.

Practice your presentation until you're satisfied with it. Then practice pointing to the appropriate places on your display board as you refer to titles, IV, DV, controls, constants, data, repeated trials, and so on. Tape or videotape yourself. Listen to or watch the tape. Think of ways to improve your presentation. Redo it.

Figure 15.4 contains a set of questions the judges might ask you. Study them. Prepare yourself to answer them. About now you are ready to perform before a live audience. Get your Mom, Dad, Aunt or Uncle, Grandparent, or if you are really desperate your brother or sister, if you have one, to listen to you. Remember you are trying to explain what you did and its importance so don't be afraid to sound excited or to show your enthusiasm for your project. You spent a lot of time doing it, now convince your audience that you did a great

Presenting Your Experiment 187

job. After you have mastered your oral presentation, have your home audience ask you the questions the judges might ask you. Tell your audience to ask them in any order as that is what judges do. Practice your responses before your live studio audience.

If the competition you enter is your state's Junior Academy of Science or one of similar format, then you will illustrate your oral presentation with overheads or slides rather than with a 3 panel display board. Typical overheads or slides include a project title page which includes Items 1-3 (unless the information in item 2 is prohibited by competition rules.) Item 4 is often just a verbal statement. Item 5, the experimental design diagram, can be on a transparency. The procedure (Item 6) may involve a listing of steps, diagrams, or slides of the experimental set-up. Item 7 may require several transparencies of data and graphs. Your conclusion (Item 8) should be a series of overheads which outline your major points, including other studies you think should be done and to whom your experiment would be important.

FIGURE 15.4 SUGGESTED QUESTIONS FOR JUDGES

Background Knowledge
Why did you decide on this topic?   What is the purpose of your project?   What library information did you find that was helpful?
Experimental Design
What was your hypothesis?   What variable did you intentionally change?   What response did you observe or measure?   What did you intentionally keep the same?   What group did you compare the others against? Why?   How many times did you repeat the experiment?
Materials and Methods
What materials did you use?   What steps did you follow in conducting the experiment?   If you had a mentor, in what ways did the mentor assist you?
Results—Conclusion
What results did you find?   How did your results relate to your original hypothesis?   What conclusion did you make?   If you conducted the experiment again, what would you do differently?   What additional experiments would you suggest?   Which groups in the community would be interested in your experiment?   What recommendations would you make to these groups?   What was the most important thing you learned from the experiment?   © *Cothron, Giese, & Rezba.* Students and Research. *Kendall/Hunt. 1989, 1993, 2000.*

Searching the Web

For more information on Presenting Your Experiment, use these search words or phrases:

"oral presentation"
"science fair displays"
"science fair rules"
"science fair judges"

to find websites such as this:
twingroves.district96.k12.il.us/ScienceInternet/OralPresentation.html

Glossary

axis—the horizontal or vertical line found at the bottom and left side of a graph; plural is axes.

bar graph—a pictorial display of a set of data using bars to indicate the value, amount, or size of the dependent variable for each level of the independent variable tested. Usually the taller the bar, the greater the value of the dependent variable.

bibliography—a list of all books, papers, journal articles, and communications cited or used in the preparation of a report or scientific research paper.

call number—the set of numbers and letters used in a library to identify, catalog, and shelve each book.

conclusion—the last section of a report of an experiment; it states the purpose, major findings, hypothesis, a comparison of the findings of this experiment with other experiments and recommendations for further study.

constants—those factors in an experiment that are kept the same and not allowed to change or vary.

continuous data—measurement or count data for which all possible values exist whether tested or not. Time and volume are examples of continuous data. When both the independent and dependent variables result in continuous data, the data can be graphed as *either* a line or a bar graph.

control—the part of an experiment that serves as a standard of comparison. A control is used to detect the effects of factors that should be kept constant, but which vary. The control may be a "no treatment" control or an "experimenter selected" control.

counts—data stating the number of items, for example, the number of bees attracted to sugar water or the number of people responding to a noise.

data—the bits of information (measurements, observations, or counts) gathered in an experiment; data takes a plural form verb, "the data *are*; the data *were*."

data table—a chart to organize and display the data collected in an experiment.

dependent variable (responding variable)—the factor or variable that may change as a result of changes purposely made in the independent variable.

derived quantity—information or values determined by calculations using collected data; examples are means, medians, modes, and ranges.

discrete (discontinuous) data—data that exists in categories that are separate and do not overlap such as brands of products and kinds of papers. When displayed by a scale on a graph, the points between the defined categories do not have any meaning. Discrete data can be graphed as a bar graph but *not* a line graph.

experiment—a test of a hypothesis. It determines if purposely changing the independent variable does indeed change the dependent variable as predicted.

experimental design diagram—a graphic illustration of an experiment

Title:
Hypothesis:

IV:					← independent variable
					← levels of the IV tested and control
					← number of repeated trials

DV: ← dependent variables
C: ← constants

experimenter selected control—the set of trials conducted for a single level of the independent variable that is selected by the experimenter to be the standard of comparison. For example, in an experiment to determine which brand of gasoline is best a "no treatment" control, using no gasoline in a test car does not make sense.

Four Question Strategy—an approach for generating a series of experiments from a topic, demonstration, or other prompt. Brainstorming response to the four questions in the strategy results in many potential independent variables and dependent variables to select among. It also results in descriptions of ways to describe or measure the dependent variable and identify many constants.

frequency distribution—a summary or graph of the amount of variation (spread) within a set of qualitative data (observations); a frequency distribution states the number of items in each category, for example, 2 red, 10 pink, and 25 green tomatoes.

graph—a pictorial display of a set of data.

horizontal (X) axis—is the line along the bottom of a graph on which the scale for the independent variable is placed.

hypothesis—a prediction of the relationship of an independent and dependent variable to be tested in an experiment; it predicts the effect that the changes purposely made in the independent variable will have on the dependent variable. Plural is hypotheses.

independent variable (manipulated variable)—the variable that is changed on purpose by the experimenter.

intervals—the equal size values represented by the equal spaces marked along the axis of a graph, or the spaces between units of a measuring device.

introduction—a paragraph at the beginning of a report of an experiment that states why the experiment was done (reason); what was expected to be learned (purpose) by doing it, and the hypothesis tested.

levels of the independent variable—the specific values (kinds, sizes, or amounts) of the independent variable that are tested in an experiment.

line graph—a pictorial display of data that can be drawn when the data for both variables are continuous data. The line in a line graph shows the relationship between the independent and dependent variables.

line of best fit—a smooth line drawn so that the totals of the distances between the line and the points above and below it are equal (roughly half the data points are above and half are below the line). The line that can be either a straight line or a smooth curve shows the relationship between the independent and dependent variable.

manipulated variable—see independent variable.

mean—the most central or typical value in a set of quantitative data; the formula for calculating the mean is:

$$\text{Mean} = \frac{\text{the sum of all the measurements (or counts)}}{\text{total number of measurements (or counts)}}$$

measurements—data collected using a measuring instrument with a standard scale.

median—the central value in a set of data ranked from highest to lowest. Half the data are above it and half are below.

mode—the most typical or central value of a set of qualitative data. It is the value that occurs most often in the set.

negative association—the inverse relationship that exists when increasing the independent variable results in a decrease of the dependent variable.

no treatment control—a control that receives none of the independent variable, for example in an experiment testing the effect of varying the amount of a fertilizer on plant growth, a no treatment control would be a set of plants that receives no fertilizer.

note card—a card used to record notes about the information found in a reference source or interview. Reference documentation is also put on each note card.

observations—data that are descriptions of qualities such as shape, color, and gender.

positive association—the relationship that exists when increasing the independent variable results in an increase in the dependent variable.

procedure—a sequence of precisely stated steps that describe how an experiment was done, including the materials and equipment used.

qualitative data—verbal descriptions or information gathered using scales without equal intervals or zero points. Such scales are non-standard scales.

quantitative data—information (data) gathered from counts or measurements using scales having equal sized intervals and a zero value. Such scales are standard scales.

range—a measure of how a set of measurements or count data is spread out. It is calculated by subtracting the minimum value from the maximum value.

reference documentation—the information needed to identify each source used in doing and reporting an experiment; this includes the author's name, the titles of the articles and or books, newspapers, or journals, as well as the date, city and state of publication, and the publisher; or, in the case of an interview, the name of the person interviewed and the time, date, and place of the interview.

reference style manual—a book that states rules for writing the reference information for the books, interviews, encyclopedias, magazines, journals, and newspaper articles used in doing and reporting an experiment.

repeated trials—the number of times that a level of the independent variable is tested in an experiment or the number of objects or organisms tested at each level of the independent variable.

responding variable—see depending variable.

results—a section of the report of an experiment that includes the data tables, graphs, and sentences that summarize any trends found in the data.

scale—a series of equal intervals and values placed on each axis of a graph.

standard scale—a scale of measurement that has both a defined zero point and equal intervals. Examples are distance scales in inches, centimeters; temperature scales in degrees Celsius or Fahrenheit.

title—a statement describing an experiment or data table. Titles are often written in the form, "The Effect of Changes in the Independent Variable on the Dependent Variable" (all important words in a title are capitalized).

trend—the general direction or pattern of the data; it is usually illustrated on a graph as a line of best fit.

types of data—the kinds of information collected in an experiment; typical types are measurements, counts, or observations.

value—the size, amount, or extent of a property described by a piece of data, count, or observation.

variable—things or factors that can be assigned or have different values in an experiment.

vertical (Y) axis—the line drawn on the left side of a graph on which the scale for the dependent variable is placed.

Appendix A

Using Style Manuals

A bibliography at the end of a research paper typically lists all the written materials that were used in the preparation of the paper. Very specific rules must be followed when preparing a bibliography. These rules are found in books called style manuals.

Most English teachers in middle and high schools use the format recommended by the Modern Language Association (MLA). In preparing the bibliography for your research paper or science fair project, use the bibliographic format taught in your school. If you intend to enter your paper or project in a competitive event, such as a science fair or your state's junior academy of science (JAS), be sure to read their rules. Some competitive events may require the use of a bibliographic format different from what is taught in your school.

The style of the Modern Language Association (MLA) was used for references throughout this book. Other styles include the Chicago Manual of Style, the American Psychological Association (APA) publication manual, and Turabian's manual for writers of term papers.

Information you would list on most books in your bibliography consist of only three parts.

- Name of the author. Give the last name first to make alphabetizing easy.

- Title of the book. Underline the title, and capitalize it according to the rules of the style manual you choose to use.

- Publication information. Include the place of publication, the publisher, and the latest copyright information.

Using the style recommended by the MLA, the bibliographic information on a book would look like this:

>Gibaldi, Joseph, and Walter S. Achtert. <u>MLA Handbook for Writers of Research Papers</u>. 6th ed. New York: The Modern Language Association of America, 2003.

If the APA style was used to reference this same book, it would look like this:

>Gibaldi, J., & Achtert, W.S. (2003). <u>MLA handbook for writers of research papers</u>. (6th ed.). New York: The Modern Language Association of America.

Compare these two styles. Can you find at least three ways they are different?

◆◆◆

Among the differences you may have noticed are: the order of the authors' names, the abbreviation of first names, the placement of the date, and the capitalization of the words in the book title.

In addition to books, other written materials you might use in preparing your project or paper may include articles in magazines or journals, in newspapers, or in encyclopedias. The basic patterns of the information you will need for different types of written materials follow. Be sure to use the specific rules of the style manual you use.

Book title: Basic pattern

>Author. <u>Book Title</u>. Place of publication: Publisher, year of publication.

Article in a magazine or journal: Basic pattern

>Author (if known). "Article Title." <u>Name of magazine</u> Month year: pages.

Article in a newspaper: Basic pattern

>Author (if known). "Article Title." <u>Name of newspaper</u> Day Month year.

Article in an encyclopedia: Basic pattern

>Author (if given). "Article Title," <u>Name of encyclopedia</u>. Date.

Citing Internet resources: Basic Pattern

Last name, First. Title [Online] Available at http://www...., date(s).

- List the author's name, title of the page, and the URL
- Indicate the date you visited the page and the last date the page was updated, if given

Example:

Pioneer Thinking Newsletter. Making Handmade Paper in 10 Easy Steps. [Online] Available at http://www.pioneerthinking.com/making_paper.html, December 10, 2003.

References (in MLA style)

American Psychological Association. Publication Manual of the American Psychological Association. 5th. ed. Washington, DC: American Psychological Association, 2001.

The Chicago Manual of Style: For Authors, Editors, and Copywriters. 15th ed. Chicago: University of Chicago Press, 2003.

Gibaldi, Joseph, and Walter S. Achtert. MLA Handbook for Writers of Research Papers. New York: The Modern Language Association of America, 2003.

Turabian, Kate L. A Manual for Writers of Term Papers, Theses, and Dissertations. 6th ed. Chicago: University of Chicago Press, 1996.

Appendix B

Conducting Interviews

There are many people in your community who would be delighted to help you learn more about science. Don't hesitate to talk about your project with people. If they can't help you, they may know someone who could provide information. Talk to your teachers and as many of your friends and classmates as you can. Your parents know many adults and may be able to suggest people who might be helpful.

People who may help can be found at universities and in industries. Depending on your topic you might contact zoos, nature centers, or museums. Check the yellow pages for medical offices, hospitals, and businesses related to your topic. Try greenhouses if your topic is plants, or electricians and engineers if you are studying some aspect of electricity. Most phone directories also have blue pages that list government agencies that might be able to provide information on your project.

You may also call an agency that you think may be able to help. However, before you call, make a list of questions you want to ask. Put each question on a separate card. Have a pencil and paper ready to take notes on the cards. You might practice conducting an interview with someone in your family first, using the following tips.

- Introduce yourself and provide information about your school and grade.

- Let them know what topic you are investigating for a science project and that you would like to ask them a few questions.

- Describe your research problem.

- Ask your questions over the phone or ask for an appointment to discuss your project.

- When taking notes, record important phrases and key words. Tape record the interview if the person agrees.

- Ask the person for suggestions of resources to read or other people to contact for information.

- After your interview, immediately review your notes and add additional information while you still accurately remember details.

Write a follow-up thank you letter. Most people who took the time to talk to you are interested in what you are doing. Keep them informed and invite them to view your finish project.

If the person could not help you, ask them for other persons to contact. Don't feel badly if some people won't talk to you about your project. Some people just won't have the time or interest in your topic.

Appendix C

Expanding an Introduction

The introduction sets the stage for a science project. It introduces the topic to readers and helps them understand what the project is about. An introduction often communicates only a few major ideas such as the reason, purpose, and hypothesis for the experiment. If the reader knows about the variables you are you are studying, this may be enough. Many readers, however, would like an introduction with more information—one that gives some background information on the organisms, types of matter, forms of energy, or processes used in the experiment. At a more advanced level, readers also need descriptions of experiments previously conducted by researchers.

When you select your variables, you probably will work with a combination of living organisms, matter, or energy. In some cases you may work with a process, such as the best method for watering plants or controlling erosion. The following questions will help you identify some important information about your variables to locate and include in the introduction of your report.

Organisms—Plants, Animals, and Other Organisms

1. Provide some general information about the organism.

    a) What is the common name of the organism? The scientific name?

    b) How is the organism classified, that is, to what kingdom, phylum, and so on does it belong?

    c) What is the organism's habitat? Where does it live? What are its habitat requirements?

2. Describe the general appearance of the organism. Provide more detailed information about the body part or system you are investigating.

    ◆ For plants, important systems to describe include the roots, stems, leaves, and flowers.

- For animals, important systems to describe include the skeletal, muscular, circulatory, nervous, digestive, respiratory, excretory, reproductive, and so on.

- Remember you do not have to describe every part in detail, just the major parts involved in your study.

3. How does the organism function? For example, how does it:

    a) obtain needed materials such as air, water, food, and so on?

    b) move?

    c) eliminate waste?

    d) respond to stimuli?

4. How do you expect the organism to act or respond in your experiment? Why?

Forms of Matter—Chemicals

1. Provide some general information about the chemical.

    a) What is the common name of the substance? The chemical name?

    b) How is the substance classified chemically?

    - Is it an element, compound, or mixture?

    - Is it an inorganic or organic substance?

    - What type of inorganic substance is it? Examples include metals, nonmetals, and metalloids or acids, bases, and salts.

    - What type of organic substance is it? Examples include hydrocarbons, starch, protein, fat.

2. Where is the substance found? Does it occur in nature? Is it manufactured?

3. What are the chemical and physical properties of the substance?

4. How is the substance used?

5. How do you expect the matter to act or respond in your experiment? Why?

Forms of Energy

1. What category of energy are you investigating? Examples include mechanical, heat, light, sound, electrical, magnetic, nuclear.

    a) How is the form of energy produced?

    b) What are some examples of the form of energy?

    c) What are some uses?

2. How is the form of energy:

    a) transferred or transmitted from place to place?

    b) changed or transformed into other types of energy?

    c) changed by interacting with matter? What happens to the matter as the energy interacts with it?

3. How do you expect the form of energy to act or respond in your experiment? Why?

Process

1. What is the purpose of the process?

2. Provide a brief description of the process.

3. Where does the process occur? It is natural? Created by humans?

4. How will you use the process in your experiment? What do you expect to happen? Why?

Combinations Abound

You can combine the questions in many different ways to locate and outline the material you will need for an introduction. For example, if you were investigating the effect of salt on the regeneration of planaria, you would use the questions for matter and organisms (animals). If you studied the effect of electromagnets on the rooting of elodea, you would need the questions for forms of energy and organisms (plants). Even a combination of questions on processes and plants work, as in the study of "The Effect of Crowding on Bush Bean Growth," described in Chapter 12. When you write your introduction, think of it as three parts:

◆ Background information on the independent variable;

◆ Background information on the dependent variable;

◆ Closing paragraph that communicates the reason, purpose, and hypothesis for the study.